UNDERSTANDING EVOLUTION OF MAN
An Introduction to Palaeoanthropology

UNDERSTANDING EVOLUTION OF MAN

An Introduction to Palaeoanthropology

P.K. Seth

KALPAZ PUBLICATIONS
DELHI-110 052

UNDERSTANDING EVOLUTION OF MAN
An Introduction to Palaeoanthropology
(Anthropology, Biology)

Rs. 890

ISBN: 81-7835-106-4

Published in 2003 in India by
Kalpaz Publications
C-30, Satyawati Nagar
Delhi-110 052
E-mail: kalpaz@hotmail.com

Lasertypesetting by: Varsha Graphics, Delhi
Printed at: Mehra Offset Printers, Delhi

Contents

Preface

A noteworthy sign of the times has been a growing interest in prehistory among people concerned predominantly with our origin. As such, the remote past has a wider scope and significance. Findings about life in prehistoric times bear on problems in psychology, psychiatry, animal behaviour, medicine and research along these lines bears on problems in archaeology and anthropology.

As a student of anthropology myself, I was never comfortable with palaeoanthropology. And at the very beginning of my career as a teacher at the University of Delhi, Late Dr. A. Sharma persuaded me to share a part of his Master's teaching load in palaeoanthropology. Very hesitatingly (for he was a very strict but a wonderful teacher), I accepted his request. It was then that I decided to work on this project to bring out a comprehensive treatise, which would inculcate interest in students in palaeoanthropology. Most of the information contained herein arose out of my lectures to students on palaeoanthropology at the beginning of their studies in anthropology over the last three and a half decades.

It is only during the past few years that a subtle turn toward origin of mankind became the central point in primate evolutionary biology. In presenting and co-ordinating material from diverse sources, I received help from many individuals. I was fortunate in having Late Dr. A.K. Mitra as my teacher at earlier stages in Anthropology. It is not only what I learnt from him but also what he conveyed simply by being himself. For the sheer flow of stimulating ideas and a glimpse of future trends, I owe a great deal to my teacher Late Dr. A. Sharma, perhaps the most articulate and forceful teacher of palaeoanthropology of our times. I hope that the facts and information recorded here will be found useful. We know so much about man and yet understand so little. I would be contended if this book encourages, even slightly, others to try to understand more.

I also received other kinds of help. Most of my graduate students prepared line drawings for me, procured literature and kept me posted with latest references in the field. I am also thankful to many of my colleagues for discussions in developing concepts in the evolutionary arena. But for their help, the scenario would have been different.

My wife, Swadesh and son Nitin provided me encouragement, stimuli and inspiration during this assignment. Nitin, in particular, helped and assisted me in preparing the illustrations, in overcoming problems with the computer, etc. Lastly, I wish to thank all who helped me during the various stages of this assignment.

July 2002 **P.K. Seth**

List of Figures

List of Tables

1

Fossilisation

Palaeoanthropology, an intriguing discipline, is the study of human evolution. It is the science dedicated to the exploring of our roots and is based on our understanding of the evolutionary process and the fossil evidence for the past events that exemplify it. Among the biological sciences, it is uniquely concerned with a single family, rather than broad inter-specific comparisons. By observing the direction of evolutionary change to modern man, the characteristics of fossil man were ticked off in a simple linear manner. Interestingly, another characteristic of palaeoanthropology is that it changes rapidly. Admittedly, much of this is imposed reluctantly by the appearance of inconvenient fossils and dates, but shifts in the explanatory framework have also taken place recently. Fossil hominids are being interpreted in the ecological and evolutionary context in which they lived.

In order to understand human evolution in its entirety, a broad base of information is required. The task of recovering and interpreting all clues left by early hominids is the work of a palaeoanthropologist. As such, it is a multidisciplinary pursuit seeking to reconstruct every possible bit of information concerning dating, structure, behaviour, and ecology of our hominid ancestors.

The events that led to the emergence of living people began far in the past. In just the last few years, the study of early humans has marshalled the skills of many diverse kinds of specialists drawn from many other sciences in developing hypotheses about the course, pattern, and the causation of human evolution. Included primarily in this exciting adventure/investigation are the geologists, archaeologists, physical anthropologists and palaeoecologists. Today's survivors provide detailed information about relationships between the living —

the relationships of today are based on the past, and can help reconstruct it. There are records left to us from prehistory, such as the preserved fossilized remains of once living organisms, humanly modified artifacts, that reflect some aspects of human behaviour, and provide evidence of the habitats in which they died. The fossilized remains of humans, their ancestors, and their relatives, are the most direct evidence for human evolution, but its study relies on many other disciplines. It draws extensively on geochronology, stratigraphy, archaeology, evolutionary and molecular biology, ecology, functional anatomy, taphonomy and a host of other sciences. For instance, palaeontologists help find fossil beds containing faunal remains, where conditions are favourable for bone preservation of pigs, elephants etc and thereby also favourable for the hominid remains. Besides, palaeontologists provide quick estimates of the approximate age of the fossil sites without waiting for the time-consuming chronometric analysis.

It is an indisputable fact that the fossil hominid record compares poorly with those of many other mammalian groups, despite the money and effort that has been invested into palaeoanthropological fieldwork in the last three decades. The hominid fossil record is really too poor to contribute significantly to the understanding of the tempo and mode of evolution, in general. Collecting and studying human skeletal remains, both in the museums and laboratories, is presently a complex, and controversial subject. Emphasis in palaeoanthropology has now shifted much more towards that of evolutionary biology as a whole, towards recognition of the ecological factors and away from a dependence upon assumptions about the uniqueness of humans and consequently of their evolutionary history. Interest in human origins and the study of human evolution has a long history.

Within anthropology, archaeologists have always acknowledged the wealth of information that can be recovered from mortuary contexts. For example, a human burial contains more anthropological information per cubic meter of deposits than any other type of archaeological feature. Beside, it represents the latent images of biological and cultural person frozen in a clearly delimited segment of space and time (Peebles, 1977).

Fossils are studied in a variety of contexts: comparisons

are made of their morphology (form), to better understand phylogeny (the genealogy of species) and to try to reconstruct the relationship between form and function in the once living organisms.

Today, through osteological analysis, our knowledge of previous cultures and human populations is greatly increasing. Information and/or the data thus acquired have historical as well as medical significance to modern humans. Therefore, the studies of the human bone are invaluable to explaining and understanding the past. From the study of bone, one can understand the nutritional status, diet, disease processes, factors affecting mortality and life expectancy, biological responses to environmental stressors and aging of our ancestors. Also thus gained is the wealth of information about the prehistoric cultural practices and social structure, population movements and about tribal origins.

Patterns of Social Organisation

Physical anthropologists and archaeologists can unravel the mysteries of the patterns of political and social organization of past cultures from a study of the bones. This is, however, greatly dependant upon data about age and sex of each burial.

Even evidence of status and marriage patterns is often visible in cemetery populations. By analysing genetic differences among men and women in the same cemetery, it is often possible to reconstruct marriage and residence patterns. By studying skeletons for indications of growth disturbances and disease, sometimes it is possible to tell whether the greater height of high status people was due to better diet and more resource, or whether they were genetically predisposed to be taller.

Study of the prehistoric human remains provides insight into the many facets of the lives of the individuals and the communities of the past cultures. Skeletal studies provide information about the lifestyles as well. For instance, well-developed muscles leave their mark on the bone and suggest heavy physical activity during life.

Ethnicity can often be determined by observing the characteristics of the facial skeleton. Statistical studies of tooth, skull and face shape can even help distinguish closely related groups within the same major ethnic group.

The death of an organism is the beginning of a creative process of biological and physical transformation: some remains are immediately recycled into other organisms or gradually converted into other raw materials for other biological structures, some remains may be removed from this system to be preserved as fossils, and some remains that escape recycling through fossilisation achieve a certain degree of immortality. A small portion of these latter remains becomes the fossil material, which serves as evidence for organic evolution through geologic time. The complex processes that preserve organic remains in rocks also leave their own traces, adding another dimension of information to fossil samples.

Taphonomy

With increasing understanding of the complex effects of preservation, these fossils offer a great deal of information about their evolutionary history, clues about their environment, and interaction with other species; all such information being coded in a complex way. Problems involving the effects of preservation on the fossil record extend through all the different phyla, back to the earliest life on Earth. The vagaries of the fossil record have a significant effect on the way we reconstruct the biological attributes of the fossil remains as seen in the original biota. Taphonomy, the study of natural processes of preservation and destruction, consists of successive stages of transformation — death, decomposition, burial and fossilisation. Taphonomic studies reveal that deposits of fossilized bone are rarely a random sample of the once living organisms.

As the fossils represent past organisms, evolutionary patterns and important biological events in the history of the earth, i.e. the long-term history of the relatively short-term (biological) processes, it is necessary to work back through the stages of fossilisation, considering possible biasing effects on fossil material at each stage. The original morphology, population size, life history, behaviour and habitat of the organisms likewise effects their potential for becoming fossilized. Interestingly, the best fossils or complete skeletal material has resulted from natural catastrophe of life burials, for instance, the preservation of human remains by volcanic eruption such as the one at Pompeii, or severe storms, epidemics, variations in the temperature, etc. Mass mortality or instantaneous death and burial create the optimal initial conditions for fossilisation.

There is little doubt that the amount of time that may be represented within and between fossil samples has significant effects on the interpretations of the evolutionary patterns as "punctuated" or "gradual": continuous or attritional fossils indicate gradual change whereas instant samples show punctuated or sudden change.

Fossils are generally regarded as 'having turned to stone' by the replacement of the original hard parts with secondary minerals. The process of mineralization, or diagenesis, is extremely varied. It is a function of the original composition of the organism together with the chemical components of the burial environment. For instance, in highly acidic tropical forest soils, the hard and soft parts of the organism can dissolve completely; organisms buried in highly compressible mud or peat may reveal less useful information than those buried in sand because of greater distortion in the more compactable sediment. For example, the Miocene *Ramapithecus* in China (8.5 million years old) was found in lignite (coal) deposits resulting in so much flattening that much of the internal cranial anatomy is inaccessible, at least by the existing methods. Likewise, it is difficult to estimate the cranial capacity in skulls that have crushed or flattened. In contrast, the skull of robust *Australopithecus* (approximately 1.5 million years old) from Kenya retained much of its original structure as it was filled with sand during fossilisation.

Since the process of fossilisation depends on local soil conditions, its' speed and the details of its progress can vary greatly from place to place. In one area, fossilisation may achieve significant results in only a few thousand years, while in another notable mineral replacement may take millions of years. Thus, the age of a fossil cannot be determined from how mineralised it has become. Fossilisation is rare and the conditions surrounding death and burial are usually so traumatic that only parts of the original organism become fossilized. Yet, the fossils themselves, interpreted in an evolutionary and ecological framework, provide the primary data for the study of human evolution.

Dating Fossils

Palaeoanthropology not only relates to discovering fossil remains but also seeks to date these materials, understand the fossilisation process, the ecological context of the fossils and in understanding the interface between biotic, social, psychological and cultural existence of man on this planet earth, i.e. human evolution in its entirety.

In an attempt to study extinct societies, it is essential to consider the ecological conditions that prevailed at the relevant time and place. This is possible only by reconstituting the climate and soil that together formed the habitat or by recovering the vegetation and animals that constitute the biome and provide man in the most literal sense with the basis of his subsistence.

Despite the paucity of fossils, only the fossil remains provide the basic evidence for interpreting evolutionary history of man. However, the importance of any fossil lies in the fact whether the fossil itself or the deposit from which it was recovered can be dated.

Dating of fossils is essential because of two basic reasons: firstly, the osteological evidence can then be interpreted phylogenetically; and secondly, the environmental information of a site could be correlated with the fossil remains. Usually when considering the age of a fossil, answers to two important questions are sought namely what is the relationship of the fossil material to the geological, floral, faunal and archaeological sequence at the site? i.e., whether the remains are contemporaneous with the stratum in which they lie, and therefore whether the site's faunal, climatic, and archaeological information can be properly associated with the find(s). However, in this respect, it is important that the layer has not been unnaturally disturbed. And secondly, what is the chronological age of the specimen in years B.P. (before present)? An answer to this query provides the chronometric age of the fossil.

Methods

Direct Methods. Dating methods would apply the dating techniques to the objects themselves, which would be the preferred method. However, fossils and artefacts are too precious to risk destroying in the process of dating. Indirect methods represent the practical approach. In this case, dating the associated finds provides the age for the fossil material or the artefact. An age can even be assigned based on the evolutionary stage of the non-human fossils associated with them - faunal correlation.

Indirect Methods. The most common indirect dating method is to date the stratigraphic layers that lie below and above the find in question. The stratigraphic layers accumulate from bottom upwards, i.e. the upper layers are the youngest and the lower layers are the oldest. Absolute dating methods depend upon some aspects of radioisotope activity.

The various dating methods, majority of which are relative dating methods, vary in precision, and allow a tentative placing of a particular object or assemblage in a time sequence. There is a great need for an internationally acceptable dating. Basically, the dating methods can be classified as relative and absolute or some combination of these. Relative dating merely establishes a sequence of events and places a specimen, event or deposit in relation to an established sequence. It provides a chronological sequence of youngest to the oldest fossil.

Chronometric Dating. Chronometric dating makes it possible to measure the actual age of a specimen, i.e. absolute dating, and provides age estimates in numerical figures. This method provides the ultimate structure, on which fossil interpretations rest and thus provides the ultimate evolutionary network (Table 1.1). An absolute dating provides the duration of time that has elapsed since the event occurred.

TABLE 1.1
Major Chronometric Dating Methods

Time Period	*Dating Method*
Modern period to 2500 B.C.	Historical documents, dendrochronology, Glacial varve
Recent times to 50,000 years approx.	Radiocarbon, Amino acid racemization
50,000 years ago to 500,000 years ago	No accurate dating technique available
500,000 years ago to the age of earth	Potassium-Argon, Fission-track

TABLE 1.2
Radioactive Elements Used in Dating

Parent Element	*Decay Product*	*Half-Life in Years*
Uranium 238	Lead 206	4.51 billion
Uranium 235	Lead 207	710 million
Thorium 232	Lead 208	13.9 billion
Rubidium 87	Strontium 87	50 billion
Potassium 40	Argon 40	1.33 billion
Potassium 40	Calcium 40	1.33 billion
Carbon 14	Nitrogen 14	5,730 years

Half-life. Radioactive decay series involve reactions presumed to occur at rates that are independent of temperature, pressure, and particular combinations of atoms as molecules. Because of these attributes, there is a constant rate of loss of the initial material and the accumulation of the end products and, therefore, a constant proportion of the two sets of elements (Table 1.2). Because there is a constant proportion of the starting materials and the end products with respect to time, such decay series may be used to determine the time elapsed or the age of a sample containing the elements of the decay series. Although the decay rates of elements vary from microseconds to billion of years, one can determine the half-life of an element as the time it takes for half the amount of starting material to be changed to the end products. Because, it is important to know only the proportion of the starting and end product substances, the absolute amounts of these materials are not essential for the calculations.

Relative Dating Procedures

This dating method is based on the assumption that the archaeological/osteological evidences are contemporaneous with the deposits in which they lie. These methods help to determine the stratigraphy at the site, i.e. the relative vertical position (age) of the objects in the soil. In an undisturbed stratum, remains in the lowest levels must be older than those lying above. And in the absence of a stratified deposit, i.e., in the event when levels are disturbed, other methods, though less conclusive, are used.

Relative dating by typology is based upon the fact that over time, manufactured objects undergo stylistic changes that reflect general cultural changes. On this basis, it is possible to arrange objects of different ages in a time sequence. This method is useful in dating surface finds.

An important method of relative time placement of archaeological remains is cross dating, i.e. establishing a similarity between an assemblage, or of some significant remains in it, and others with a previously determined age in another locality. Cross dating ties a site into a pre-existing scheme. For instance, when a site is excavated and the remains analysed, these are cross-dated with remains from other sites to establish relationships. A major means of relative dating involves testing of bone in order to establish its contemporaneity with other bones in a stratum, or to the soil deposit itself.

Stratigraphy. The first relative dating method commonly known as stratigraphy is based upon the law of superposition meaning that the lower layer (stratum) is covered by another layer. In other words, stratigraphic dating is the scientific study and description of the order and succession, i.e. age, of the earth's crust formed of sedimentary rock strata.

The major limitation of this method, however, is the fact that the earth's various disturbances may shift about the various layers of rock or the objects buried in them. This would make it difficult to reconstruct the sequence of events therein.

Three major analytical methods of relative dating include fluorine dating, nitrogen dating and uranium dating. These methods are especially important in disturbed sites where it is necessary to determine the contemporaneity of the materials. In these methods, the relative age of a fossil is determined by comparing its chemical composition with other fossil bones of known ages from the same site or the same area if they are preserved under comparable conditions.

Fluorine Dating. Fluorine dating employed frequently in dating of bones, depends upon ground water seepage through the bones. Fluorine in the water combines with the bone's calcium to form a compound known as fluorapatite, the measurement of which — from various bones in a site — determines their contemporaneity or lack thereof. This method allows bones of an assemblage to be tested for stratigraphical equivalence since contemporaneous bones should contain roughly the same amount of fluorapatite. Uncovering of Piltdown hoax was only possible because of this technique. Fluorine test is mainly devised for the purpose of checking the age-relation between bones and the deposit in which they have been found.

Nitrogen Dating. Nitrogen dating of the bone is often coupled with fluorine dating for the results complement one another. Bones that have accumulated little fluorine retain much nitrogen and vice-versa. The method is most useful for assaying whether enough collagen degradation has occurred in an attempt to date the bone directly by radiocarbon method, which entails destroying part of the sample (already often small). The Galley Hill and Piltdown material was first subjected to fluorine and nitrogen tests before the bones were finally radiocarbon dated.

Uranium Dating. Uranium dating involves the principal that the longer a bone lies in a deposit or a stratum the more uranium it absorbs. Since uranium is radioactive, its content within the bone can be measured by counting the amount of radioactive deterioration. The major significance of this method is that it does not involve destruction of the bone as in the fluorine and nitrogen methods.

Absolute Dating Techniques

Radiocarbon Dating. Radiocarbon/C–14: All living matter contains a small but practically constant proportion of the radioactive isotope of carbon, C–14, which is produced by cosmic-ray bombardment of nitrogen atoms in the outer atmosphere. On death, the radioactive carbon in its tissues is no longer replenished from the atmosphere, and it disintegrates at a constant rate – the quantity is halved in 5,570 years. Thus, by measuring the radioactivity of the carbon extracted from an ancient specimen of organic material, its age can be calculated.

The time t since the death of the organism would be given by:

$$t = I/\lambda \log_e I_0/I$$

where λ is the decay constant of radiocarbon, I is measured activity of the ancient sample, and I_0 is the measured activity of modern organic material.

Limitations: The backward limit of the C14 method is between 60,000 and 70,000 years. And a certain minimum quantity of organic carbon must be available in the specimen or sample. Thus, calcined bone is undateable whereas charred bone is potentially dateable.

Obsidian Dating. This procedure is based upon the fact that a freshly made surface of obsidian, a variety of volcanic glass, will absorb water from surroundings to form a measurable hydration layer. This layer is not visible to the naked eye and should not be mixed up with the patina that develops on many materials as a result of alteration or chemical weathering. The measurement of the death penetration of hydration on the artefact should, then, provide a measure of the time that elapsed since the manufacture of the artefact. Hydrated obsidian has a higher refractive index than does the non-hydrated obsidian. This can be determined under a microscope.

Fission-Track Technique. This technique dates material ranging from 20 years to 1,000,000,000 years before the present, the controlling factor being the presence of uranium impurities, usually in trace amounts such as are commonly found in nature. This method has advantages over C14 and K-A methods: use of radioactive decay from 70,000 years ago back to nearly the beginning of the Pleistocene epoch, 2-3 million years ago. For ages greater than 70,000 years, the short half-life (5,660 years) of C-14 makes carbon dating inapplicable; and although Potassium-Argon dates in the range of 70,000 to 3,000,000 years, the long half-life (1,300 years) of K-40 makes measuring such ages experimentally laborious, for great care must be paid to the selection of suitable materials and to their processing.

Fission-track dating is based upon the fact that over geological time spontaneous fission of uranium-238 impurities produces minute, sub-microscopic damage trails in most insulating solids, both long periods of time. As the damage sites are permanent, a simple count of their number allows an age to be measured. The density of track depends on the uranium content as well as the age of the sample (i.e., the 'fossil' tracks counted), viz. exposes the sample to a known number of low energy (thermal) neutrons, which induce fission of a small fraction of the uranium atoms in the sample. The sample is immersed in the proper chemical reagent, which allows rapid attack of the narrow regions of damage and slower dissolution of the adjacent unchanged material. Such treatments are used to create etch pits of optical size, each one making a single fission site. The method requires sufficient uranium to produce a track density, which can be counted within reasonable time.

Potassium-Argon Dating. This method can be applied particularly for the dating of lower geological ages by virtue of the abundance of potassium and the improvements in high vacuum technique and sensitivity for measuring very small quantities of rare gases. The natural potassium comprises of 93.2% K^{39}, 6.8% K^{41}, and 0.0118% radioactive K^{40}. At the time of the formation of earth, the abundance of K^{40} was about 0.2%. Most of the isotope K^{40} decayed into its daughter isotopes Ca^{40} and A^{40}; this method determines the ratio of these two isotopes in a mass spectrophotometer. Since the high half-life, of K^{40}, $1.30 \pm 0.04.10^9$ y, is comparable to the age of the earth, a small amount of K^{40} is still present, whereas the nuclide palladium 107 and iodine 129, with shorter half-lives, have decayed some billion years ago. For each 100 K^{40} atoms, which decay, 89% become Ca^{40} by β (beta) decay and 11% become A^{40} by K-capture.

The half-life of potassium^{40}, $1.30.10^9$ y, is long enough for appreciable A^{40} to be formed in potassium minerals within geologically interesting periods. By measuring the radiogenic A^{40} – concentrations and the total content of potassium of a mineral, one may determine the time, which is necessary to produce the radiogenic A^{40} – concentration by radioactive decay:

$$A^{40}/K^{40} = (R/1+R).(e^{\lambda t} - I)$$

where R = branching ratio; λ = decay constant. This period is designated as the potassium-argon age of the sample.

The content of potassium is usually determined by a flame photometer using small concentrations of isotopic dilution analysis or neutron activation analysis. The determination of concentration of argon is made in three steps: melting of the samples for the extraction of gases, purification of the extracted rare gases and the determination of the quantities by an isotopic analysis (using mass-spectrophotometer). In addition to the radiogenic A^{40}, the material to be dated also contains argon from the atmosphere with an isotopic ratio A^{40}:A^{38}: A^{36} = 296:0.19: 1. During measurements, atmospheric argon is added from small leaks, from diffusion through walls of the apparatus, etc.

Material Used

Suitable material like potassium-bearing rocks, such as igneous or once molten forms like lava and tuffs; sedimentary rocks, such as glauconite or clays.

Period

The half-life of K^{40} is very long (1.3 billion years). In practice, samples less than 0.5 million year cannot be dated unless they are very rich in potassium (such as sanidine and leucite). In theory, potassium-rich samples as young as 5,000 years can be dated with this method. The maximum period that can be dated by this method is the age of the earth.

Thermoluminescence (Tl)

This method measures the number of electrons caught up in defects in the lattice structure of the crystals. Dates for burnt flints found at the cave sites of Kebara and Qafzeh in Israel support the idea that the *Neanderthals* were not the direct ancestors of modern humans, because they arrived in the Middle East several tens of thousands of years after the first anatomically modern people in the area (TL method).

Neanderthal Dating techniques/Mass Spectrophotometric U-Series techniques date sites beyond the range of radiocarbon dating. For instance, the Middle Palaeolithic sites of Israel contain both the *Neanderthals* at Tabun and Kebara, and the anatomically modern hominids at Skhul and Qafzeh. This has given rise to conflicting views of their relationship in Levant. As the U-uptake by dental material is site-specific, the new High Precision Mass-Spectrophotometric U-Series ($^{230}Th/^{234}U$) technique reveals that the early modern *Homo sapiens* were probably coeval with and locally pre-dated *Neanderthals* in the Levant (McDermott *et al.,* 1993).

Palaeomagnetism Technique

This method of dating fossils, an important means of

crosschecking, is based on the constantly shifting nature of the earth's magnetic field, both in direction and intensity.

This method involves the meticulously obtaining samples of sediments that contain magnetically charged particles. It is well established that these particles maintain the magnetic orientation they had when they were consolidated into the rock and thus provide a kind of 'fossil-compass'. The palaeomagnetic sequence is then compared against the potassium-argon dates to check, if they agree. Major epochs of palaeomagnetic time are:

Normal (ethnocentrically, northern orientation)
←0.7 mya
Reversed (southern orientation)
←2.5 mya
Normal
←3.4 mya
Reversed

Electron Spin Resonance Technique

Relatively new dating technique of Electron Spin Resonance detects the trapped electrons in-situ without heating. Presently, this method is used for tooth enamel alone.

Thermoluminescence and *Electron Spin Resonance* techniques both date materials between a few thousand to 1 mya. Application of these methods revealed that the modern human fossils in the Middle East were at least 1,00,000 years old and not 50,000 years as inferred by other methods. These new dates reveal that the *Neanderthals* of the region dated at 60,000 years old could not be ancestral to the modern humans as thought earlier.

Faunal Correlation Technique (Biostratigraphy)

Another dating method based on the evolutionary changes in well known groups of mammals, called *Faunal Correlation* or *Biostratigraphy* provides yet another means of cross checking dates obtained by other methods. Because of regular changes within an evolving lineage, approximate ages can be extrapolated from areas where dates are known to other less well-known areas by recording, which genera and species are present.

Dendrochronology

Dating by tree rings and Varve Chronology (annual glacial deposit) are other absolute dating methods that do not involve radioactive elements. Although neither of these dating methods have a direct bearing on dating early human fossils, yet they are both ingenious techniques with important regional applications.

Amino Acid Racemization Technique

A chronometric dating method rather recently applied to human fossils is the *Amino Acid Racemization*. This is based on the fact that all living organisms have L-amino acids in their proteins and after death, and over long periods of time, all these L-amino acids except glycine undergo change called racemization and become non protein D-amino acids. The proportion of these D-amino acids increases with time. Thus, the age of the skeletal material can be estimated by determining the amount of change, racemization that has occurred.

2

Primate Radiation

Record of man's early evolution is still fragmentary. Man, even today, is regarded as the most evolved among the primates. The fossil evidence suggests that the earliest primates were small, forest dwelling, and insectivorous mammals not larger than a rat. Primate adaptive radiation began sixty-five to seventy million years ago in the Palaeocene. Many lines of this radiation have survived till today: both the processes of anagenesis and cladogenesis being responsible for this radiation. Though the living primates could be arranged in order of increasing anatomical and behavioural complexity, modern primates are the end products of their own evolutionary lines. The presence of similarities between fossils and living forms suggest that the successively evolving populations grade into each other and form a trend.

The origin of primates can be traced to the late Cretaceous (van Valen and Sloan, 1965). Each primate radiation probably occurred when the earth was undergoing extensive changes: proliferation and an enormous increase in the variety of ivies, shrubs and trees. With opportunism as an outstanding characteristic of the mammalian evolution (Simpson, 1949), primates, therefore, developed grasping extremities, nails instead of claws on their digits (except tree shrews, marmosets and tamarins), forwardly directed eyes, diminished olfactory sense and possess the ability to sit, stand and walk upright. These major skeletal changes/adjustments were necessary for adaptation to tree dwelling leading to a shift in the dietary preference and habitat exploration. As a result, major modifications took place in their skull, face, hands and feet.

Primate Development

Early Tertiary Period

At the beginning of the geologic phase, now called the Tertiary period – about 66.4 million years ago – there were in existence the most primitive of the primates. So primitive are their anatomical characters (as shown by fossils) that it might be impossible to determine that they were primates but for the fact that they mark a gradation towards more highly organized creatures that definitely come within the category of primates.

The climate of the early Tertiary period was warm, with wide tropical and subtropical zones extending from the equator up to the higher latitudes in both the Old and the New World. During the Palaeocene epoch, which lasted for about 8.6 million years (c.66.4 million-c.57.8 million years ago), there were many primates in existence; 60 genera have been recognized and grouped into eight families. Three of these families had long chisel-shaped teeth that resembled those of the rodents with which they competed for a similar ecological niche, or habitat.

Early in the Tertiary period, during the Palaeocene and Eocene epochs (from about 66.4 million to about 36.6 million years ago), more advanced primates appeared that belonged to the same zoological groups as the modern lemurs and tarsiers. A widespread family, the Adapidae, divided into two subfamilies, the Adapinae and the Notharactinae, represents the Lemuriformes in the fossil record. The Tarsiiformes are known from one family, the Omomyidae, containing four subfamilies: the Anaptomorphidae, the Omomyinae, the Ekgmowechashalinae, and the Microchoerinae. It would appear that the characteristic tarsioid (tarsier-like) specialization of the skull and hind limb were already well advanced in the fossil forms that are known, but some of the European genera have some structures that indicate relationships with the early monkeys. Generally speaking, however, little is known of the Eocene ancestors of the Old World monkeys and apes; thus, the Eocene epoch terminated after about 30 million years of primate evolution with lemur-like forms and tarsier-like forms but having little or no evidence of anything else.

Explosive primate radiation took place in the Palaeocene and the Eocene epochs and it dwindled in the Oligocene (oligo =

small). Lemuroids, Tarsioids and platyrrhines have been recognized in the Palaeocene and Eocene epochs of America, Europe, Egypt and Burma. The New World witnessed the birth of three-fourth of the primates and their development and origin of man probably took place in the Old World. *Amphipithecus*, implying both ways an ape, a platyrrhini, found in the Eocene of Burma is considered ancestral to *Parapithecus* (*Para* = near) of Egypt.

Later, in the Oligocene epoch (36.6 to 23.7 million years ago), which followed, there came into existence primitive monkeys and exceedingly primitive anthropoid apes. Excavations into the Fayum deposits of Egypt, of Oligocene age, have disclosed a deltaic shoreline bounded by tropical bush country that supported an extensive fauna including rodents, hyraxes, pigs, small elephants, and primates.

Fayum deposits in Egypt, of the Oligocene epoch resulted in fossils of *Propliopithecus* (*Pro* = before; *plio* = more), the earliest anthropoid ape (of the size of a cat) on record and *Parapithecus* (known from some lower jaws (thirty million years ago), a very small sized (squirrel-like) and earliest Old World Monkey on record having a generalized Tarsioid appearance. Gregory (1957) regarded *Propliopithecus* as a primitive gibbon and *Parapithecus* as a primitive monkey. These fossils are characterized by traits generally essential for adaptation to arboreal life: grasping extremities, nails instead of claws, pentadactyly, an opposable thumb and big toe, forearm consisting of ulna and radius, reduction of snout, forwardly directed eyes, closure of orbits from behind, and an enlargement of the visual centers.

Apidium, which is included in the same family, the Parapithecidae, is also found there. These may be the forerunners of African monkeys. Also derived from the Egyptian Oligocene are several fossil apes of primitive type; these include *Aeolopithecus*, which may be an ancestral gibbon, and *Aegyptopithecus*, which may be ancestral to the modern great apes. One other fossil ape from the Fayum that deserves special mention is *Propliopithecus*, formerly believed to be an ancestral gibbon. It has been suggested, primarily on the basis of its generalized dentition, that *Propliopithecus* is possibly ancestral to the hominids. This however is a speculation that requires much more evidence before it can be generally accepted.

Primate fossils from Oligocene time are rare. The remains of two apes, *Propliopithecus* and *Aegyptopithecus*, were found in the desert of Egypt. They had small brains, and long snouts. Their skulls were like those of monkeys or lemurs and their teeth like those of modern apes. They lived in trees and had tails.

Miocene Epoch

The Miocene epoch began about 23.7 million years ago and lasted about 18.4 million years. It was a remarkable phase in primate evolution in which there appears to have been an increase in the number of larger primates that were widely spread throughout the Old World, including Europe, Asia, and Africa. The large number of fossils recovered from widely separated sites over a long period of time has led to taxonomic confusion and more than 50 species have been described and classified in 20 genera; this number has been considerably reduced by an overdue taxonomic revision. The large Miocene hominoids appear to belong to three groups, the *Sivapithecus*, the *Dryopithecus*, and the *Proconsul* groups. Within these groups are some 24 species from Europe, Africa, and Asia.

The earliest ape-like form, *Parapithecus*, is known from Oligocene deposits in Egypt, which also have yielded remains of several kinds of Old World monkeys. The earliest New World monkeys are known from late Oligocene deposits in South America. During the Oligocene, South America was isolated from Central and North America, and a unique mammalian fauna developed there. Remarkably, many South American mammals of the Oligocene exhibit extreme parallelism in adaptation to forms that are found elsewhere in the world and are not even closely related.

The Miocene fossils of Asian and African apes suggest that the Asian apes formed a distinct and diversified group in Asia and Southeast Europe by fourteen million years ago (e.g., *Sivapithecus, Ramapithecus*, etc). Significant other fossils have been reported from the Miocene of Europe, Egypt and Africa. The lower Miocene epoch is often referred to as that of the 'age of apes'. Skeletal remains from Africa were classified into three major groups: *Proconsul* (a non-specialised ape), *Sivapithecus* (a transitional form between the anthropoids and man), and *Limnopithecus* (an early type of gibbon).

From Europe, the fossil material of *Pliopithecus* was recovered. The name implies that the individual is considered ancestral to the modern gibbons. A small sized primitive gibbon (*Prohylobate*) frequented Egypt during the Miocene. This ape was slightly bigger in size than the *Propliopithecus*. This epoch, thus, saw the rise of the generalized apes of large size but regarded as the off shoots of *Propliopithecus* of Oligocene.

Dryopithecids (a very heterogeneous group) and *Oreopithecus* were occupying India and Europe during the Pliocene. In fact, the dryopithecinae fossils represent the most interesting and controversial part of the primate evolution. Pliocene marks the decline, both in numbers and diversification, of these closely related Miocene forms and the beginning of the primitive hominids.

The *Dryopithecus* group includes the first specimen, found in 1856 in Saint-Gaudens, Fr. (*Dryopithecus fontani*), whose cusps and fissure pattern, known as the Y-5 arrangement, is typical of dryopithecines. Other examples are now known from Hungary (*Rudapithecus*), Spain (*Dryopithecus laietanus*), and China (*Dryopithecus eiyuanesis*).

The *Proconsul* groups arrived in Africa during the earlier part of the Miocene period. It includes three species – *Proconsul africanus*, *Proconsul nyanzae*, and *Proconsul major* – as well as *Rangwapithecus gordoni* and several other smaller-bodied apes. The second group of African and Middle Miocene apes is the Oreopithecidae, which includes *Nyanzapithecus*, the large *Afropithecus*, and two species of *Kenyapithecus*, all from East Africa.

Parapithecus

This is undoubtedly the most primitive known catarrhine monkey from the lower Oligocene in Fayum, southwest of Cairo in Egypt. The fossil remains first reported and described by Schlosser (1910) consist of a well-preserved complete lower jaw (Fig. 2.1) with all the teeth present (Werth, 1919): this is the earliest known anthropoid ape, on record, from the same Fayum deposits which gave us *Propliopithecus*.

The small-sized lower jaw is of the pointed type and the simian shelf is absent therein. The two halves of the mandible of *Parapithecus* converge at an angle of 33° (a very primitive condition among the primates) while this angle in the gibbon's ranges from 16 to 21; it being 0 in Man, i.e., the two rows are almost parallel. The dentition conforms to the catarrhine dental formula of 2.1.2.3. This suggests that this earliest known Old World Monkey had a generalized tarsioid appearance. Schlosser mistook the larger lateral incisors as canines (as they appear more pointed than the preserved canines) and opined that *Parapithecus* had retained the prosimian dental formula of 1.1.3.3. If this formula were correct, then this condition indicates that this fossil had affinities with the tarsioids. This contention is supported by the marked convergence of the dental arches anteriorly. The body size of *Parapithecus* is like that of a big squirrel or a modern living tarsier.

Fig. 2.1. *Parapithecus* Mandible

Gates (1948) suggests that *Parapithecus* stands in between the Eocene Lemuroids (for instance, Anaptomorphidae of north America) and *Propliopithecus* (comparatively larger/earliest

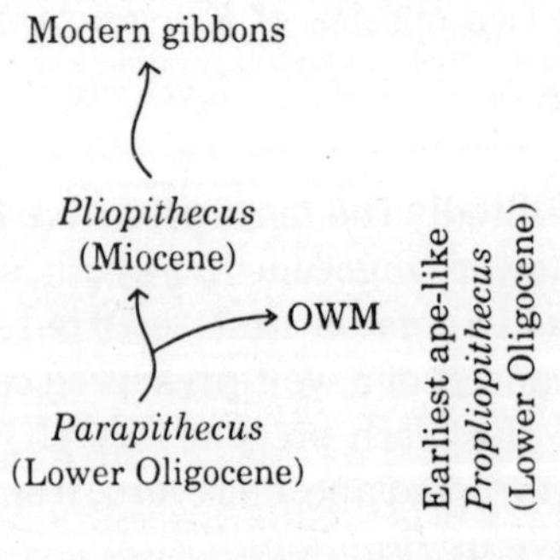

Fig. 2.2. Fossil Lemuroids of North America (Palaeocene and Eocene)

ape). This is more advanced and in turn also leads to the Miocene *Pliopithecus* and the recent Hylobatidae (Fig. 2.2). He regards *Parapithecus* as an ancestor of the Old World Monkeys as well. This is also in conformity with the vertical ramus of the mandible (Clark, 1959; Fig. 2.1).

At one time, Gregory even suggested that it was 'a generalized super-tarsioid' or 'pre-monkey' of the Old World'. Recently, Gregory, like Montagu and Gates, regards *Parapithecus* as an intermediate form between the lower primates (e.g., lemuroids and tarsioids) and the catarrhines (whole Old World Monkey groups including even apes and man): all other known members of the OWMs having become diversely specialized away from the primitive *Parapithecus* type (Fig. 2.3). Thus, Gregory has climbed down to almost the same opinion as was earlier held by Montagu (1945).

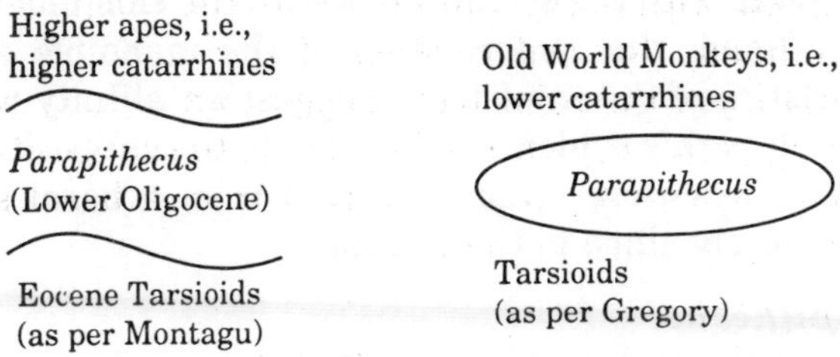

Fig. 2.3. Status of *Parapithecus*

Montagu (1945) considers that the three premolars have been reduced to two premolars yielding thereby a formula of 2.1.2.3 for the first time. The morphology of the teeth of *Parapithecus* indicates that they did have a mixed diet: unspecialised canines; blunt premolars and molars suggest that they were not carnivorous; incisors lean forwards pointing to a mixed diet. According to Gregory (1957), *Parapithecus* stands at or is the very line of descent leading (Fig. 2.4) to the anthropoid apes and eventually to man.

Parapithecus → Great apes → Man

Fig. 2.4. Line of Descent – *Parapithecus* to Man

Sir W.E. Le Gros Clark maintains that the jaw of *Parapithecus* certainly suggests that the Hominoidea may have been derived directly from the tarsioid ancestry, i.e., independently of the Cercopithecoidea. They may also be considered belonging to the basal representatives of the cercopithecoid stock preceding the stage of bilophodont specializations but may well be extremely primitive hominoids having affinities with the Hylobatinae.

The status of *Parapithecus* being regarded as a primitive Oligocene hominoid has often been questioned for although the molars quite closely resemble those of the early fossil apes, *Propliopithecus* and *Pliopithecus*; their cusp pattern is of a generalized type not uncommonly found in other early mammals such as condylarths. However, a consideration of their dentition (especially the premolars and molars) shows that *Parapithecus* was a primate with closest affinities with the Hominoidea. On the other hand, the conformation of the mandible and the characteristics of the front teeth suggest an affinity with the tarsioids. It is this evidence, which led to the supposition that the Hominoidea were directly derived from a basal stock of primates closely allied to the tarsioids.

Propliopithecus

This is the earliest known anthropoid ape found from the Fayum deposits of Egypt (Schlosser, 1911) and belongs to the Oligocene. The finds of this very primitive cat-sized anthropoid comprise of a lower jaw with all the teeth but with its ascending ramii missing. The remains were discovered from the same Fayum deposits, dried uplake beds, from which the mandible and teeth of the earliest known Old World Monkey, *Parapithecus* were found.

The tooth row is three-fifth of the size of that in the living gibbons (30 mms in length compared to 51 mms in the gibbons). *Propliopithecus* jaw is prognathus, relatively short, deep and more pointed than in the gibbons. Its molars resemble those of *Pliopithecus* and *Parapithecus*. As a result, *Propliopithecus* is more advanced and leads to the Miocene *Pliopithecus* (upper Miocene and lower Pliocene) (gibbanoid), mainly based on its dental morphology, and to the Hylobatinae of today (gibbons). This indicates that the dentition of a generalized hominoid type had already become established.

These species are relatively small to medium-sized early catarrhines characterized especially by the bulbous-cusped, bunodont molars, relatively deep mandibular corpus and nearly upright 'ramus'. The dental and cranial adaptations of *Propliopithecus* also compare favourably with those of the platyrrhines.

Propliopithecus is referred to as a 'very primitive gibbon' and thought to have descended from a primitive and generalized catarrhine monkey much like *Parapithecus* (Gregory, 1957). Buettner-Janusch believes that the lower molars of *Propliopithecus* are like those of the pongids – this may be a primitive hominoid pattern. This anthropoid ape constitutes a better ancestor for apes than man than does *Parapithecus* is the view of Montagu (1945). Others consider it to be more primitive than the lower Miocene. Schlosser (1911) regards *Propliopithecus* to be closely related to gibbons (*Pliopithecus*) from the middle Miocene deposits at Sansan. He thinks that this may possibly be a descendant from a species of the *Anaptomorphus* group, which passed through a Cebian stage-giving rise directly to *Pliopithecus* from which both the anthropoids and Man may have been derived.

Limnopithecus

Hopwood discovered *Limnopithecus* in the year 1933 from the early Miocene formations of Kenya in East Africa. The finds of this early type of gibbon consist of several fragments of the mandibles, teeth and the limb bones. The dentition of *Limnopithecus* conforms to the hylobatinae pattern while its relatively unspecialised limbs show many resemblances to those of the arboreal quadrupedal monkeys.

The limb bones show a combination of cercopithecoid and pongid (compare with modern gibbons) features. It has been recognised that the spider monkey (*Ateles*), as result of the convergent evolution, shows a number of anatomical characters resembling those of gibbons, but in less extreme form. Their arms are shorter than those of the modern gibbons whereas the legs are quite strongly developed. It seems probable that this archaic gibbon-like *Limnopithecus* gave rise to *Pliopithecus*, which spread over Europe and Asia in the late Miocene and in early Pliocene. And those of the individuals who reached the Far East in the Pliocene probably gave rise to the modern living gibbons (Clark, 1959).

Hooton (1954) argues that *Limnopithecus* may well have developed from the Fayum *Propliopithecus* in a generally gibbanoid direction. Boule and Vallois (1957) consider that this fossil presents characteristics of an archaic gibbon-like *Pliopithecus*, and is probably derived from *Propliopithecus*. Certain of its features of development indicate that *Limnopithecus* developed along an independent branch.

Pliopithecus

This fossil ape is represented by a number of excellently preserved fossils from late Miocene to early Pliocene of Egypt and Europe. These finds are reminiscent of gibbons but are more similar in detail to *Propliopithecus*, *Dendropithecus* and *Dryopithecus*. *Pliopithecus* is known from a number of European populations, most of which have been given species or even subgenus rank on minimal criteria. The fossil material consists of a complete postcranial skeleton (from Neudorf in former Czechoslovakia), mandible (Fig. 2.5) and portions of skull. The other definite records of the anthropoid apes are from the Miocene of the Vienna basin in Austria. There are numerous fragmentary fossils named *Pliopithecus* (Fig. 2.5) recorded by Hoffman (1893).

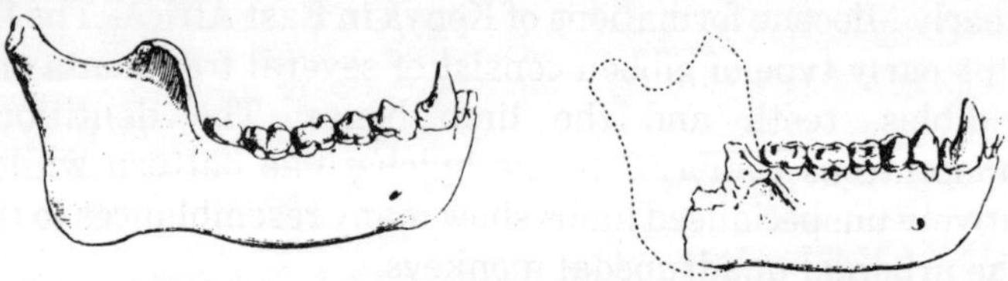

Fig. 2.5. Mandibles of *Hylobate* (left) and *Pliopithecus* (right)

There seems to be close affinity between *Pliopithecus* and Pongidae based on some important dental characteristics:

The greater relative breadth of the anterior teeth, all premolars and the first molar; the longer mandibular symphysis – a general developmental trend evident in many mammals and has less affected the gibbons than the gorillas. This also indicates the presence of a more primitive condition; and the greater divergence of the mandibular ramii.

The mandible is 'V' – shaped posteriorly (a prosimian influence) and has a trace of the simian shelf. *Pliopithecus* dentition is similar to that of *Limnopithecus* and its mandible is like that of the modern siamang – an animal of the same body size; the modern gibbon has lost all these traits.

The sabre-like canine teeth are absent and the jaw is deeper than that in the modern gibbons. *Pliopithecus* is often referred to as a rather more advanced member of the hylobatinae group. The nasal region, which is wider than in modern gibbons, again shows prosimian affinities. The facial skeleton has a greater interorbital distance. Its body is gibbon-like in details of the pelvis and the vertebrae, sternum is broad and flat as in the brachiators and man, and the clavicle is 'S'-shaped as in orang-utans and chimpanzees rather than bowed as in modern gibbons. *Pliopithecus* is a primate holding its trunk by far erect but is not extremely brachiating like gibbons.

The newly discovered material of *Pliopithecus* from Vindovonian deposits of Austria are very similar, both in dentition and the postcranial structure, to *Limnopithecus*, an early type of gibbon, from East Africa; though in some features, this fossil age is less specialized (Clark, 1959). The characteristic great length of the forelimb in the living gibbons is not present in these Miocene fossils. For instance, the intermembral index is 95 in *Pliopithecus* compared to 165 in living gibbons.

The lower limbs are as long as that in siamangs. The shoulder and elbow joints in *Pliopithecus* are not as swivel-formed as in the modern animals; the radius and ulna articulate with the carpal bones as in the prosimians and other monkeys; the tarsal bones are longer, the metatarsals and toe bones are shorter than in the siamang; and similarly are the hand bones.

The differences in the limb bones of *Pliopithecus* from those of the modern gibbons show prosimian resemblances more than the living cercopithecoid monkeys. In the general plan and details of its cranial anatomy and dentition, *Pliopithecus* was ancestral to the modern gibbons (Gregory, 1951) and not distantly related to *Proconsul* and the Miocene ancestors of the chimpanzee. In short, *Pliopithecus* was not yet a full time brachiator or altogether an arboreal quadruped. In this respect, *Pliopithecus* resembles man and the chimpanzee.

Zapfe (1958) argues that the teeth, narrow snout, widely set eyes and some features of the collarbones and other long bones in *Pliopithecus* are gibbon-like or anthropoid-like. While many other skeletal features including the vertebral column, upper limbs and the general body proportions are those of the monkeys. Some of these features are distinctly primitive resembling those in the prosimians; the limb bones are in particular astonishingly primitive. Simons suggests that on morphological grounds, the primitive gibbon-like genera *Pliopithecus* and *Limnopithecus* are indistinguishable. Hooton (1954) believes that *Pliopithecus* probably is a descendant of *Propliopithecus*, who is an almost full-fledged gibbon.

Prohylobates. The name itself suggests that this fossil material is gibbon-like. *Prohylobates* were discovered from Moghara on the northern edge of the Libyan Desert and are so gibbanoid that they can be classified under the family Hominidae.

Dryopithecinae

These fossil remains represent the most interesting and the controversial part of the primate evolution. The entire subfamily Dryopithecinae is named after a mandible (*Dryopithecus fontani*) found by Lartet in 1856 in France. They have been reported from China, northern India, Africa and parts of Europe, viz. France, Germany, Spain. The dryopithecines, a very heterogeneous group as they are, represent a stage of primate evolution rather than a single phylum and its branches. The fragmentary nature of the fossils renders it difficult to determine clearly the derived features, which permit recognition of fossil members of this group. Broadly, speaking, three genera of dryopithecinae are recognised: *Dryopithecus*, *Sivapithecus* and *Proconsul*. Numerous species of *Dryopithecus* have been described. They show considerable variation in their dental anatomy, some suggesting an approach to the chimpanzee, some to the orangutans and yet others to the gorilla. It is often argued that different members of Ponginae, including the living great apes, *Australopithecus* and probably man as well arose from them.

This ape complex is represented by teeth, jaw fragments, cranial and long bones (humerus shaft, ulna and a femur). The humerus bone with its both ends missing was reported from France and uncertainly labelled as that belonging to

Dryopithecus fontani. A complete femur found in Germany has been assigned to *Paidopithex* (Boule and Vallois, 1957). (These long bones are gibbon-sized). Dryopithecinae fossils range in size from animals as big as gibbons to as large or larger than the modern gorillas. They are distinguished from the Hominidae on the basis of their dentition alone.

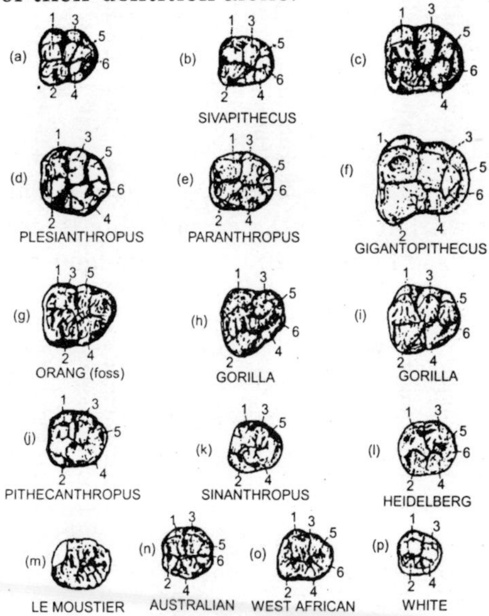

Fig. 2.6. Y-5 Patterns of the Left M_3

The incisors are small and more vertical compared to the Ponginae. Canines are larger than those in the Hominidae. The lower premolar is sectorial in shape. The molars usually increase in size within the series $M_1<M_2<M_3$. Their characteristic cusp pattern 'Y-5' is not commonly found in modern men (Fig. 2.6). This suggests that the pattern is of fossil origin. The primordial crown pattern of human lower molars is a set of three grooves in the form of a 'Y' lying on its side with its tail pointing forwards and its two arms pointing to the rear. In each obtuse angle are stationed two cusps, and in the acute angle, a fifth cusp is located – the Y-5 pattern (Coon, 1963, pp. 360). In the teeth of modern human beings, the molar crown patterns have been simplified in two ways: the groove pattern has changed from 'Y' to a simple '+' and the number of cusps has reduced from five to four or even less (three or two).

The entire Dryopithecinae complex can be subdivided into two main groups:

TABLE 2.1
Siwalik Fossils: Partial List/Details

Adapids			Gingerich and Sahni, 1979, 1984
Palaeotupaia sivalicus		Haritalyangar	Chopra and Vasishat, 1979
Indoadapis shivaii	fragments of maxillae and mandible	Haritalyangar; Middle Pliocene	Chopra and Vasishat, 1980
(probably remnants of early Tertiary Adapidae)			
Indraloris himalayensis	right jaw fragment with M_2-M_3; left jaw fragment with P_4-M_2, distal and lingual margin of M_1 damaged, M_3 partly preserved		Pilgrim, 1932; Chopra and Vasishat, 1980
Indraloris lulli	left M_2	Haritalyangar early Pliocene	Lewis, 1933; Chopra and Vasishat, 1980
Pliopithecus krishnaii	hylobatid left M_3	Nagri beds; Haritalyangar	Chopra and Kaul, 1975, 1979
Dryopithecus *P. sivalensis* (Lydekker, 1879)	both maxillae; left teeth missing; right C, 2 PMs, 3 Ms damaged and worn; large	Jabbi, now in Pakistan Dhok Pathan Level (?) of Miocene-Pliocene deposits	Pilgrim, 1915

(Contd.)

	cross-sectional area at symphysis; more chewing power		
	largest molar; size ranges between gorilla and Gigantopithecus bilaspurensis	Nagri zone, Alipur	Prasad, 1989
Dryopithecus punjabicus	posterior jaw fragment with M_2M_3 on two sides more conc in Haritalyangar, (Krynine, 1937)	Near Chinji Village	
Dryopithecus giganteus	an extremely large M_3	Nagri formation at Alipur, now in Pakistan	Pilgrim, 1915; Madden and Lewis, 1980
	resembles M_3 in *Sivapithecus* (Lewis, 1937); both sampled from similar/ identical populations; but distinct from *S. indicus*		
*Ramapithecus brevirostris**	Maxillae with 3 M of almost equal size w/o a cingulum different from *R. punjabicus*	Haritalyangar Nagri beds	Lewis, 1934; Prasad, 1964
*Ramapithecus punjabicus** Simple molar patterns, low and rounded cusps, extremely bicuspid molars		Nagri zones of Haritalyangar; Ramnagar; Chinji zone in Attock and Jhelum; Yunnan province in China	Dutta *et al.*, 1976; Simons, 1972

(Contd.)

Sugrivapithecus			Lewis, 1936
Sivapithecus indicus	Partial maxillae, part of nasal aperture	inferior Chinji formations; Middle Miocene	Raza *et al.*, 1983
		Kangra	Gupta, 1968
	Fragment of left jaw, isolated M_1, crown of last lower M	Nagri beds of Haritalyangar	Chopra and Kaul, 1975
Sivapithecus lewisi		Nagri beds of Kangra Distt	Pandey and Sastri, 1968
Sivapithecus sivalenis		Nagri beds of Haritalyangar	Chopra and Kaul, 1975
Sivasimia chinjiensis	Left M^1; resembles *Pongo pygmaeus*	Nagri beds of Haritalyangar	Chopra, 1983
Gigantopithecus bilaspurensis	hominoid mandible; almost all teeth present	Dumehra, 2 kms from Haritalyangar Dhok Pathan stage of Upper-Middle Siwaliks (Middle Pleistocene)	Chopra, 1968a,b; Simons and Chopra, 1969a,b
Gigantopithecus blackii		China; 500,000 – 1 my BP postdates the Indian variety by millions of years	von Koenigswald, 1935, 1952; Woo, 1962

* dentition more hominid than pongid like (Simons, 1961, 1964c; Simons and Pilbeam, 1965; Prasad, 1969b; von Koenigswald, 1972, 1973).

(i) Western group of Eurasiatic Dryopithecinae

- *Dryopithecus fontani* (certainly a pongid) – from middle to upper Miocene of Western Europe
- *Dryopithecus darwini* – from middle Miocene of Austria
- *Dryopithecus germanicus* – from lower Pliocene of South Germany
- *Neopithecus occidentalis* (*Sivapithecus occidentalis*)
- *Hispanopithecus laietanus*
- *Udabnopithecus garedziensis* – from the erstwhile USSR
- *Austriacopithecus* – from middle Miocene of lower Austria (limbs show both pongid and non-pongid characters)

(ii) Eastern group of Eurasiatic Dryopithecinae

- Known mainly from the Pliocene of Siwalik Hills in North India. This region, especially in India, is a highly productive radiation center of fossil pongids (Table 2.1) during the second half or late Miocene and lower Pliocene.

Dryopithecus

Dryopithecus, a genus of extinct apelike animals, is representative of a group of small, generalized apes that contains the ancestors of both the modern apes and humans. Although *Dryopithecus* has been known by a variety of names based upon fragmentary material found over a widespread area including Europe, Africa, and Asia, it appears probably that only a single genus is represented. *Dryopithecus* is found as fossils in Miocene and Pliocene deposits (23.7 to 1.6 million years old) and apparently originated in Africa.

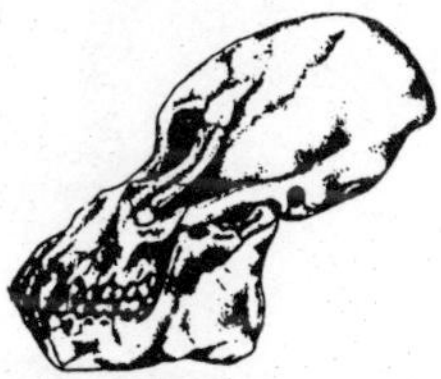

Fig. 2.7. *Proconsul africanus* skull: Lateral profile

Dryopithecus is the name given to the fossil remains of a number of ape-like creatures (about 500 to 600 individual fossils) that lived about 20 million years ago in Europe, Asia, China, former USSR, Turkey, Saudi Arabia and Africa. They persisted in the Pliocene as well. One of the well-known fossils from East Africa, *Proconsul* (Fig. 2.7), may have been the ancestor of the gorilla or the chimpanzee. *Dryopithecus*, a middle and upper Miocene European hominoid, retains a number of plesiomorphic characters thereby positioning it before the dichotomy between the African and Asian great apes (Andrews, 1992; Simons and Pilbeam, 1965; Sola and Köhler, 1993).

Remains of *Dryopithecus* have been reported both from Europe and North America for they both enjoyed a warm temperate to tropical climate during the Tertiary periods. Numerous generas and species of this fossil ape have been named mainly on the basis of bone fragments, particularly the teeth. These fossils, easily distinguished from the Hominidae on the basis of dentition alone, range in size from that of gibbons to that of the modern gorillas or larger than them. For instance, their incisors are small and more vertical than those of the Ponginae; the lower third premolar is sectorial (shearing type); molars are elongated from front to back (Fig. 2.8) and usually progressively increase in size within the series; and the limbs are of generalized nature, i.e., there is no indication of specializations in respect of brachiation or bipedalism. They are, thus, the best candidates to be the ancestors of the living African great apes.

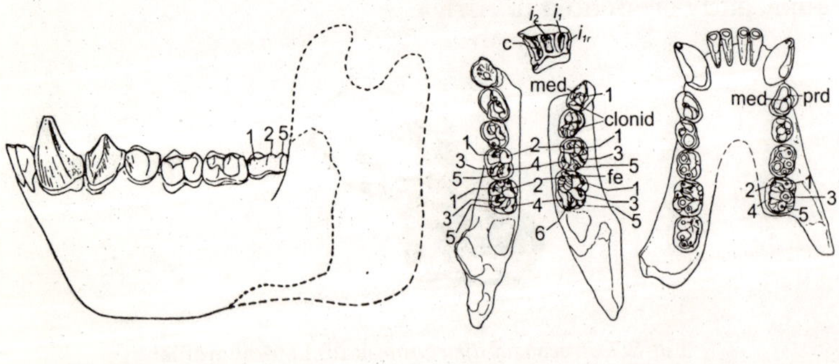

Fig. 2.8. *Dryopithecus fontani* lower jaw (adapted from Gervais and Gaudry)

The fossil remains of *Dryopithecus*, with few exceptions, are limited to teeth and pieces of jaws, i.e. comprise of incomplete fossils (Lartet, 1856; Gaudry, 1878, 1890). The exception being a humeral shaft (from France) with both ends missing and reclassified as *Dryopithecus fontani*. Another exception being a femur from Germany that was earlier labeled as *Paidopithex* (Boule and Vallois, 1957) but now assigned to the genus *Dryopithecus*. Simpson (1945) classifies this complete femur as that belonging to gibbons. Ulna, another exception, found in France has been given a date of about 13 to 9 million years ago. Amongst the most complete of the remains of any Miocene hominoid individual are the limb bones, cranial and jaw fragments, a few vertebrae and ribs from Ruisinga Island. The two sets of teeth recovered from the lower Pliocene lignite deposits/beds (brown coal intermediate between peat and bituminous coal – an excellent preservative) in Yunan in China have been assigned to *Dryopithecus keiyuanensis*. This dentition is nearly as hominoid as they look pongid, i.e., more hominoid than those from Europe.

Several distinct forms of *Dryopithecus* are known, including small, medium, and large, gorilla-sized animals. In many ways, as might be expected, *Dryopithecus* is rather generalized in structure and lacks most of the specializations that distinguish the modern apes from modern humans. The canine teeth are larger than those in humans but not as strongly developed as those in modern apes. The limbs were not excessively long – an adaptation in the apes for swinging through the trees. The skull lacked the well-developed crests and massive brow ridges found in modern apes.

Fiedler (1956) and Ehrenberg (1960) include *Proconsul*, *Dryopithecus*, *Indopithecus* and *Sivapithecus* and a few other fossil genera amongst the dryopithecinae. Dobzhansky (1962) maintains that the Dryopithecinae were certainly not identical with the modern apes and their leg bones indicate that at least some of them had not yet acquired specialisations that go with the brachiating habits. Clark (1959) believes that the diversification of pongid genera in late Miocene and early Pliocene of India suggests that this region (India) provided a secondary center of dispersal. Probably, from this center, the ancestors of modern orangutans spread eastwards to find their 'present' habitat (in Malaysia). This contention is supported by the fact that one of the middle Miocene genus *Palaeosimia*, from

India, shows striking similarities with the orangutans in their molar pattern. The dryopithecinae jaws are shorter than those of the other pongids. Their teeth are not larger than those of the living members of the genus *Homo*, i.e., certainly man-like, whereas others had large teeth. As the dryopithecinae remains, particularly from the Siwalik Pliocene deposits of Northern India, consist only of jaws and teeth, not only is the body-size of all these unknown, but also there is no indication even whether these fossil apes walked or took to brachiation. Dryopithecinae fossils represent an advanced stage of evolution than does the *Proconsul* and *Limnopithecus*. These Miocene fossils are more akin to man than any other known monkey. During the second half of the Tertiary period, the hominid primates, particularly the pongid ape stock underwent rapid evolutionary development and diversification.

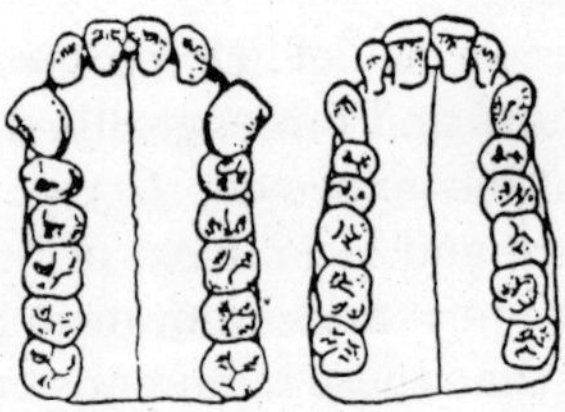

Fig. 2.9. Maxillary tooth rows of *Gorilla* (right) and *Dryopithecus* (left)

Dryopithecus gave rise to the modern gorillas and chimpanzees. Sometime during the Miocene Epoch, the dryopithecines gave rise to a derivative that eventually led to the earliest humanlike forms and finally to humans. A form close to this branching of the dryopithecine stock is represented by the genus *Ramapithecus*, distinguished by its more advanced dentition. The dryopithecines probably inhabited forested areas.

Dryopithecinae disappeared about 9 million years ago, both from Europe and East Africa, when their habitats provided less forest cover and foods had become seasonally more abundant. (A dryad is a wood nymph in Greek language and probably inhabited oak forests – a possible habitat for European dryopithecines). Leakey (1963) suggests an independent family comprising of *Proconsul*, some fossils of genus *Sivapithecus*, *Brahmapithecus* (not *Ramapithecus*) and *Sugrivapithecus* – all ape-like fossils which differ from true Pongidae in lacking a simian shelf, in the structure of their teeth, in the position of the roots of its molars – process maxillary, in the absence of turns, in the nature of their mandibular condyles and the way their teeth are arranged in the mandible.

Lartet (1856) and Gaudry (1878) believe that *Dryopithecus* is more closely akin to man than any other known monkey. On the contrary, reasons given by Gregory and co-workers (1938) and Clark and Leakey (1951) for separating *Sivapithecus* and *Dryopithecus* are inadequate. All authors agree that the lower teeth exhibit no features worthy of generic distinction. The upper teeth, though, are more strongly different than the lower dentition (Clark and Leakey, 1951). These comparisons were mainly based on '*Dryopithecus*' *punjabicus* and other isolated maxillary molars from Europe. However, '*Dryopithecus*' *punjabicus* is no longer considered a member of the dryopithecinae group and the isolated molars are inadequate for this purpose. Therefore, the reference of the Miocene and Pliocene *Dryopithecus*, and *Proconsul* genera to a single genus *Dryopithecus* seems unavoidable; the range of variation in them also does not exceed that present/encountered among species and subspecies of a single modern genera of the hominoids such as *Pan* and *Pongo* (Simons and Pilbeam, 1965).

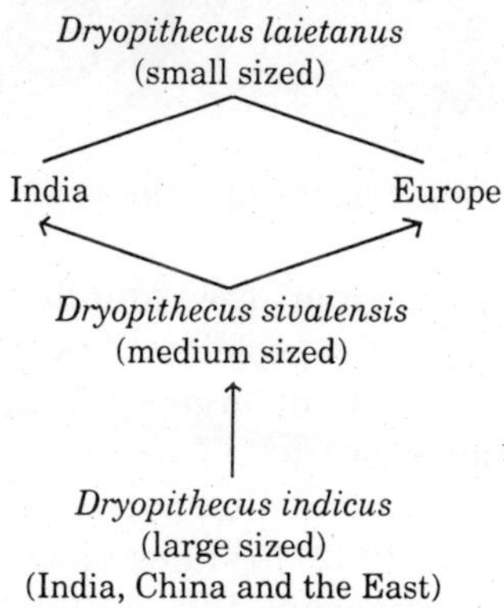

The fossil hominoid jaw from the Miocene of Namibia shares many characters with the older fossil hominoid, *Afropithecus*; the former being smaller than the latter. It is significant that the furthest southwards extension of the hominoid range in the Miocene (Namibian specimen) was phylogenetically linked with the furthest northwards extension of similar age (from Saudi Arabia) (Andrews, 1992). This mandible represents the only fossil evidence of a pre-Australopithecinae stage of the hominoid evolution in Southern Africa (Conroy *et. al.*, 1992).

Ramapithecus

The first *Ramapithecus* fossils (fragments of an upper jaw and some teeth) were discovered in 1932 in fossil deposits in the Siwalik Hills of Northern India. No significance was attached to these fossils until 1960, when Elwyn Simons of Yale University began studying them and tried to fit the jaw fragments together. Based on his observations of the shape of the jaw and of the dentition – which he thought were transitional between those of apes and humans – Simons advanced the theory that *Ramapithecus* represented the first step in the evolutionary divergence of humans from the common hominoid stock that produced modern apes and humans.

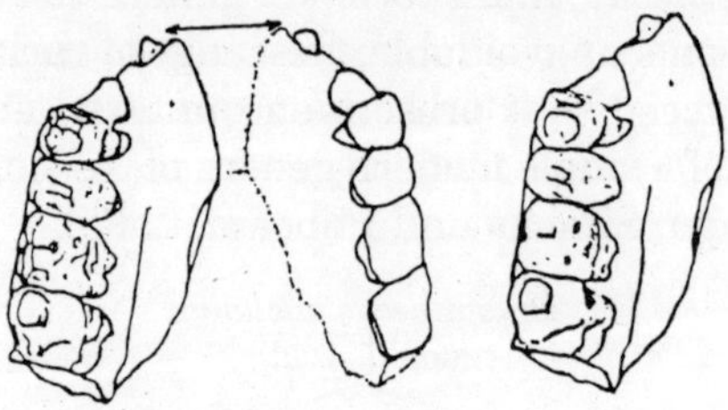

Fig. 2.10. Maxilla of *Ramapithecus*

Ramapithecus (from the Middle and Late Miocene epochs), represents the earliest known hominid and its existence establishes its divergence from that of the African apes fifteen million years ago. This basic tenet is now regarded as questionable, even if an alternative 'correct' answer cannot be provided. Lewis (1932) first described the fossil remains of a fragment of an upper jaw (Fig. 2.10) from Haritalyangar in the Siwalik Hills (India) and christened it *Ramapithecus brevirostris*. This fragment has a wide curving jaw with an arched palate (man-like feature) and was so named as it resembled the Indian God 'Lord Rama'. It has a short snout (Fig. 2.11), a feature characteristic of the apes.

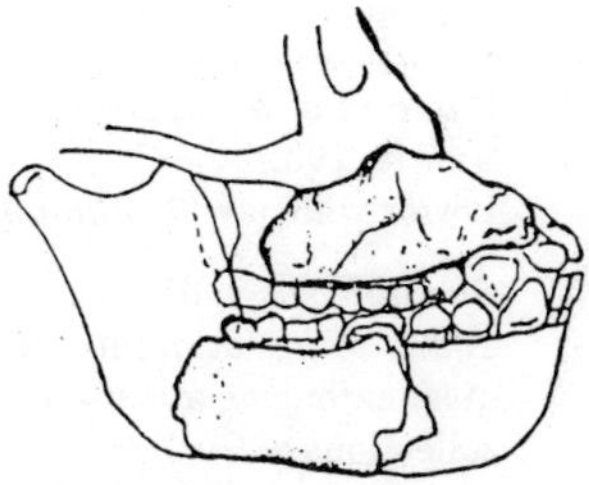

Fig. 2.11. *Ramapithecus punjabicus*

Simons (1960) re-examined the earlier finds designated as *Dryopithecus punjabicus* (Pilgrim, 1910, 1915) and opined that they were similar to *Ramapithecus brevirostris*; they were reclassified accordingly. Following this, remains of similar fossils, primarily limited to pieces of jaw bones and isolated teeth, were recovered from other sites (Table 2.2) and believed to be fourteen to seven million years ago. A revaluation of all these earlier fossil apes, given many different names, suggests that all these fossils should be classified as *Ramapithecus punjabicus* (Fig. 2.11). Even the faunal situation associated with most of these remains is broadly similar to that witnessed in the Indian sites.

Largely, dryopithecinae fossils from Europe and Asia between fifteen and ten million years ago, comprised of three major groups: *Sivapithecus*, *Gigantopithecus* and *Ramapithecus*. The former two primates had a face as large as that of modern chimpanzees and gorillas, while the latter had a small face. This latter fossil clearly shows greatest similarity with later hominids (though the former two also show some hominid characteristics). *Sivapithecus*, mainly Asian in distribution, is often classified as a dryopithecinae and is, therefore, an ape rather than a hominid.

TABLE 2.2
Ramapithecine Sites

Site	*Material / Fossil*	*Age*
Siwalik Hills of India (Haritalyangar) and Pakistan Potwar Plateau) (1910-1932)	Widely curving fragment of an upper jaw with an arched palate most man-like; *Dryopithecus punjabicus* (Pilgrim 1910, 1915)	9-12 mya

(Contd.)

(Contd.)

	Ramapithecus brevirostris (Lewis, 1932) Lower wide jaw; *Brahmapithecus*	
Lufeng basin, Shihuiba China	*Sivapithecus yuannensis* (Pilbeam, 1986) (orang-utan lineage) most extensive ape fossil collections	8 mya
Candir, Turkey (1974)	*Ramapithecus* (?) jaw	8 mya
Athens, Greece (1944), Miocene	*Graecopithecus/Ramapithecus freybergi* – lower jaw	Circa 8 mya
Hungary	*Rudapithecus hungaricus*	
Fort Ternan, Kenya (1961)	*Ramapithecus wickeri,* upper and lower jaw fragments including teeth synonymous with *Ramapithecus brevirostris*	14 mya
Germany Austria		

From amongst the various ramapithecinae finds, the *Brahmapithecus* lower wide jaw fits well with the widely curving upper jaw fragment of *Ramapithecus brevirostris* (Simon, 1961) denoting that the two might have belonged to same/similar individuals. The facial features (sloped and slightly concave facial profile) and the anterior dentition (forward jutting of canines) suggest similarities with orangutans. This clearly

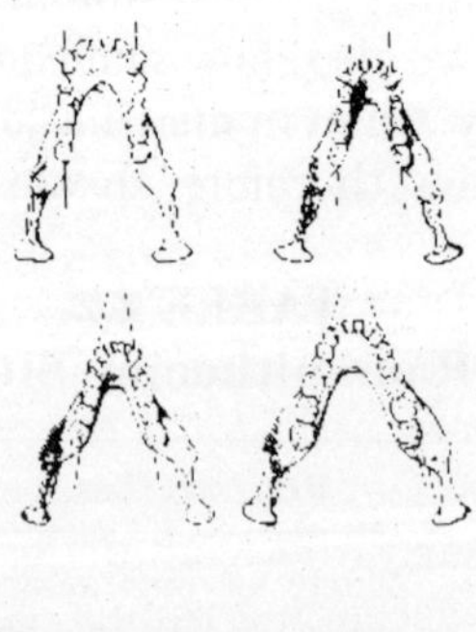

Fig. 2.12. Dental arcades of *Ramapithecus, Australopithecus, Pan* and *Gorilla* (note the divergence in the two halves)

shows that the two are the same. Of the finds of *Ramapithecus*, from the Neogene deposits of Potwar plateau in Pakistan (Siwalik Chain), the front premolars are 'still primitively single cuspid'. This is not true of the Siwalik fossils from India for they (PM_3) show distinct lingual cusp, i.e., bicuspid structure. While the Fort Ternan material displays an inner cusp, there is a transverse wear facet on the Candir material (Andrews and Tekkaya, 1980).

These man-like apes had adapted themselves to a way of life quite different from that followed by most of its forest dwelling relatives of the dryopithecinae groups. It's jaws and teeth reflect an adaptation resembling that of the australopithecines. *Ramapithecus* jaws are intermediate between *Australopithecus* and the Miocene apes in terms of the divergence of their tooth arcades (Fig. 2.12). Besides, both *Ramapithecus* and *Australopithecus* have the mandibular body, particularly in the molar region, as much thick bucco-lingually as it is deep there from top to bottom (i.e., upwards down). However, this is in contrast to the lower apes in which this region is much thinner compared to its depth. Even their lower jaws are buttressed by two shelves against the stresses of heavy chewing. These similar features – thickened and buttressed jaws, flattened molars, reduced incisors and canines – together with the proportionately higher ascending ramus in *Ramapithecus* and *Australopithecus* compared to modern man and the African apes, slightly forward shifting of the coronoid process with respect to the molars, deepened upper jaw (Fig. 2.10), thickening around the roots of the teeth, flat and short snout, and the farther forward setting of the cheek bones, show striking similarities in their adaptations. However, there is a large time interval between the two.

Upper incisors and canines tend to be directed downwards rather than forwards. Dental system of certain other mammals like hoofed herbivores, herbivorus rodents and even elephants, who mostly chew by means of a rocking or grinding action (which includes some nearly horizontal forces) have a vertical lengthening of the ascending ramus, or upward extension, at the rear of the lower jaw, which is set at a right angle. On the contrary, man and chimpanzee not only lack the thick tooth enamel and robust jaws, but also have a short and backwardly sloping ascending ramus. Ramapithecines are known to have flourished in Eurasia till around eight million years ago. Whereas the earliest evidences of *Australopithecus* and *Homo* seem to be those found in Ethiopia and Tanzania since 1974. Of these, the most ancient are all less than four million years old.

Ramapithecus lacks the post-cranial evidence and all inferences regarding its status are based largely on dental remains:

- Reduction in size and differentiation of its anterior teeth,
- Skeletal and muscular modifications permitting postural uprightness and erect bipedalism,
- Effective and adjustment to, and exploration of the terrestrial habitat,
- Increase in brain size,
- Extensive manipulation of the natural objects and development of the motor skills to facilitate tool making, and
- Carnivorous predation and addition of meat protein to diet.

When the ramapithecine fossils were being re-examined in the West, a series of similar fossils were being unearthed in the late Miocene coal fields of Shihuiba, near Lufeng in the Yunnan Province in China. This discovery was made in the year 1956 from a stratum six to seven meters thick containing many alternating layers of brown coal and fine sand. Analysis of the data obtained from these fossils demonstrates unequivocally that at least two forms, *Ramapithecus* and *Sivapithecus*, statistically quite different from each other, must have existed at Lufeng (Wu and Oxnard, 1983). At the same time, Kay and Simons (1983) argued that the Chinese material must be assigned to the genus *Sivapithecus* not withstanding the Chinese claim that these belong to two distinctly separate genera. The *Ramapithecus* finds form China show the type of general morphology suitable for the origin of the *Australopithecus* dental mechanism. Anatomy of the dentition and the face of *Ramapithecus* bears sufficient similarity to that of *Australopithecus* to justify its placement in the same family (due to convergence).

During the middle Miocene, jaws and teeth have been found in Pakistan, India, Kenya, Turkey and Greece. These have been assigned to the genus *Ramapithecus*. The thick enamel on the molars together with the relatively small incisors (forward teeth) suggests that this was a possible hominid ancestor. However, recent finds from Pakistan and Kenya have led many palaeontologists to believe that *Ramapithecus* is not after all an ancestral hominid. Instead, its hominid-like characteristics merely reflect evolutionary development, which occurred independently, several times, during the course of the anthropoid evolution. (For a time in the 1960s and 70s, *Ramapithecus* was thought to be the first direct ancestor of modern humans). As additional fossils from Europe, Africa and Asia accumulated during the 1970s, *Ramapithecus* was regarded as a small-bodied species of the genus *Sivapithecus* (Greenfield, 1979). Newly recovered cranial remains of *Sivapithecus* from Pakistan and Turkey suggest that these Asian apes probably represent the ancestry of the orangutan, and have no particular affinities with the hominids.

During the last few years, the phylogenetic status of *Ramapithecus* had become controversial with the discovery of more complete cranial/skeletal material from Pakistan and China. Some of these Miocene fossils are assigned to the pongids, especially related to orang-utans (Andrews, 1982; Pilbeam, 1982; Ward and Pilbeam, 1983; Greenfield, 1979, 1980; Kay, 1982; Ku and Lu, 1979; Xu *et al.,* 1978) while others consider them to be the earliest hominids (Chopra, 1983; Kay, 1983; Kay and Simons, 1983). Reconstruction of the biological attributes of the ramapithecine complex of fossils requires a considered evaluation of the classical anatomical methods and the recent molecular clock.

Simon's theory was strongly supported by his student David Pilbeam, and it soon gained wide acceptance among anthropologists. The age of the fossils (about 14 million years) fit well with the then prevailing notion that the ape-human split had occurred at least 15 million years ago. The first challenge to the theory came in the late 1960s from biochemist Allan Wilson and anthropologist Vincent Sarich, who at the University of California at Berkeley, had been comparing the molecular chemistry of albumins (bloods proteins) among various animal species. They concluded that the ape-human divergence must have occurred much later than the dates for *Ramapithecus* (it is now thought that the final split took place some 6 to 8 million years ago).

The hominid status of *Ramapithecus* (Jolly, 1970; Pilbeam, 1970) does not reconcile with the evolutionary time scale imposed by the "molecular clock" of Sarich and Wilson (1967). Adopting a date of thirty million years for the divergence of the apes from the Old World Monkeys, Sarich and Wilson inferred a date of five million years for a three way split among the lineages of man, gorilla and the chimpanzees. This implies that *Ramapithecus*, whose remains are dated from fourteen to seven million years ago, could not have been an ancestor of man alone. Sarich (1971) declares that 'one no longer has the option of considering a fossil specimen older than eight million years as a hominid 'no matter what it looks like'.

Anthropologists initially dismissed Wilson and Sarich's argument, but biochemical and fossil evidence mounted in favour of it. Finally, in 1976, Pilbeam discovered a complete *Ramapithecus* jaw, not far from the initial fossil find, that had a distinctive V shape and, thus, differed markedly from the parabolic shape of hominid jaws. He soon repudiated his belief in *Ramapithecus* as a human ancestor, and the theory was largely abandoned by the early 1980s. *Ramapithecus* fossils subsequently were found to resemble those of the fossil primate genus *Sivapithecus*, which is now regarded as ancestral to the orangutan; the belief also grew that *Ramapithecus* probably should be included in the *Sivapithecus* genus.

Sugrivapithecus

A 'man-like ape' from the upper stratum of the Siwalik Hills in North India was first reported by Lewis in the year 1934. These fossil remains comprised of fragments of jaw and teeth. Lewis opined that these dental characteristics 'bear a much closer resemblance to the hominians than any other known monkey'. Clark (1955) felt that 'the small size of the teeth, especially the canines, and the relatively simple construction of molars suggest the possibility that *Sugrivapithecus* was not far removed from the transitional phase which perhaps marked the evolutionary derivation of the hominid type of dentition from the characteristics of most other Pliocene and Miocene apes'. Clark (1959) further argues that 'such diversification of the pongid genera in India (and Dryopithecine in Europe) provided a secondary center of dispersal. Probably, from this center, the ancestors of modern orangutans spread eastwards to find their present habitats which the latter occupied in Malaysia'. (*Palaeosimia*, a middle Miocene genus from India, shows striking similarities in its molar patterns with the orangutans).

Sivapithecus

The *Sivapithecus* group is recognized from fossils first found in the Siwalik Hills of Pakistan and includes specimens from that region formerly known as *Ramapithecus*; other examples are known from the Candir sites and Pasalar in Turkey, Lu-feng (Lufeng) in China, as well as from Kenya. The most remarkable specimen, with almost a complete face, comes from the Potwar Plateau of Pakistan and shows many similarities to the face of modern orangutan.

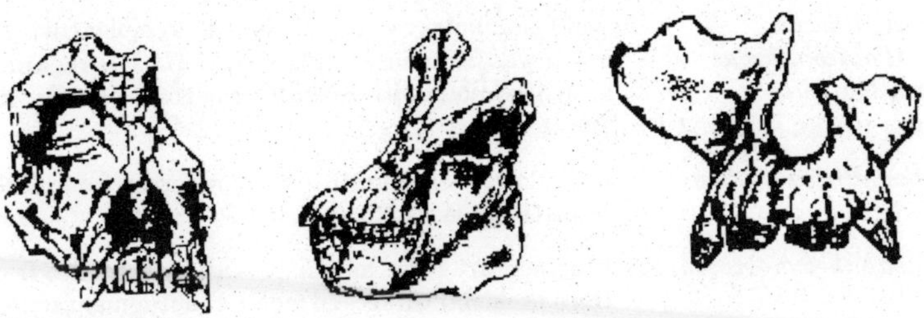

Fig. 2.13. Comparison of *Sivapithecus* and Orangutan skulls (left-Orangutan; middle-*Sivapithecus indicus*; right-*Sivapithecus meetei*)

Numerous fragments of maxillary bones, mandibles and isolated teeth are attributed to *Sivapithecus* (Table 2.5), (23.7 to 5.3 mya) a Miocene ape, from Chinji, Haritalyangar in Siwalik Hills (India), Africa (Jabi Village) and Northern Spain. Various species of this anthropoid ape have been known till date.

TABLE 2.3
***Sivapithecus* Finds**

Sivapithecus sivalensis	Highly crenulated right upper molar found near Chinji (Lewis, 1937), and shows affinities with the orangutans. It is also referred to as *Palaeosimia rugosidens*.

(Contd.)

Sivapithecus himalayensis	Ramus of the left lower jaw found from Haritalyangar (Pilgrim, 1927). Lewis (1937) assigned it to *Sivapithecus indicus.*
Sivapithecus orientalis	Right maxilla having partial/complete teeth from canine till third molar reported from the Nagri zone in Haritalyangar (Pilgrim, 1927). The alveolus for the central incisor and the root of the lateral incisor is also present. Thus, the relative size and spacing for the entire upper dentition could be estimated. This is one of the most complete dryopithecine upper dentition known. Lewis (1937) considers it to be synonymous with *Sivapithecus indicus*.
Sivapithecus middlemissi	Mandibular fragment with lower second and third molars intact discovered by C.S. Middlemissi from village Ramnagar, near Jammu (India) (Pilgrim, 1927). Lewis (1937) reclassified it to the species *Sivapithecus indicus.*
Sivapithecus (Palaeopithecus sylvaticus)	Left mandibular ramus with fourth premolar till third molar was found intact (Pilgrim, 1927). It is regarded as synonymous with *Sivapithecus sivalensis* (Lewis, 1937).
Sivapithecus indicus	Right lower second/third molar reported from north India (Pilgrim, 1910).
Sivapithecus occidentalis	Fragmentary right mandible with intact teeth from third premolar till second molar and fragmentary left mandible with first and second molars discovered in Northern Spain (Villalta and Crusafont, 1944).
Sivapithecus spp	Lewis (1934) assigned a maxillary fragment containing first and second molar to *Ramapithecus*, species *hariensis* but later on redesignated it as *Sivapithecus sivalensis* (Lewis, 1937).
Sivapithecus spp	A left mandibular ramus with crowns of third premolar till second molar, alveoli for incisors and most of the symphyseal region intact found from Haritalyangar area (Prasad, 1962). There is extreme wear on the molars suggesting that they belonged to an aged individual. The left canine was lost and its alveolus was partially resorbed and filled with cancellous bone. The material has been attributed to a new species, namely *Sivapithecus aiyengari* (Prasad, 1962).
Sivapithecus africanus	Fossil remains reported from near village Jabi in Africa (specific locality and horizon are not known)

(Contd.)

(Contd.)

	(Clark and Leakey, 1952). Macinnes (1943), following Hopwood (1933), referred all the dryopithecines from Ruisinga and Songhor to two generas, *Xenopithecus* and *Proconsul*. Later on, Macinnes regarded them as belonging to a single variable species, *Sivapithecus africanus* (Clark and Leakey, 1952); later diagnosed as *Proconsul nyanzae*. Leakey (1963) further argues that *Sivapithecus africanus* should be treated as a species with the new genus *Kenyapithecus* but one, which is more primitive and perhaps ancestral to '*wickeri*' of the lower Pliocene. The difference cited in the characters here are insufficient for discriminating *Sivapithecus africanus* from *Sivapithecus indicus* for they fall within the range of variability of the extant hominoid species (Simons and Pilbeam, 1951). Ruisinga deposits are evidently younger and the Siwalik deposits probably older, than previously thought, thereby explaining why the *Sivapithecus* remains from both these areas are so similar – *Sivapithecus africanus* is identical with *Sivapithecus sivalensis*.

According to Pilgrim (1927), *Sivapithecus* belonged to the direct progenitors of family Hominidae. But this contention is rather based on fragmentary material.

Lydekker (1879) was the first to study the fossil remains and raised doubts about the assigning of the lower jaw to *Sivapithecus* and not to *Palaeopithecus*; the upper jaw of the latter was discovered by him. He considers that the molars are more human-like while the canines are more anthropoid in appearance. Lydekker opines that the differences distinguishing man from apes usually occur mainly in the skull and other skeletal parts (postcranial) (canine teeth being tusk-like is one trait apish in nature occurring in the dentition). Lydekker classifies *Sivapithecus* among hominians while Gregory thinks it to be closely related to *Dryopithecus*. Gates (1945) maintains that *Sivapithecus* apes are very distinctly infrahuman. There is nothing to connect them with the gibbons, which are recognizable at least from the Miocene as a separate group. As parts of the skull of *Sivapithecus* have not been discovered, its status cannot be established. Lydekker argues that 'consequently its place in the classification cannot be yet established with any certainty'. Pilgrim's researches show that this ape represents, in its dentition, a sort of transitional stage between the anthropoids and Man. According to Simons and Pilbeam (1965), the range of variation among the *Dryopithecus*, *Sivapithecus*, and *Proconsul* does not exceed that present among species and subspecies of single modern generas of hominoids such as chimpanzee and gorilla. Thus, they refer the Miocene and Pliocene *Dryopithecus*, *Sivapithecus* and *Proconsul* species to a single genus, *Dryopithecus*. On the other hand, Pilgrim's (1927) restoration of the fossil material more resembles the human form than any other anthropoid jaw (living or extinct). Morphologically speaking, the fact remains that in the Miocene epoch, there were apes (with quite diverging characters = human anthropoids) roaming about in Asia.

Recently, more fossil material attributed to *Sivapithecus* has been reported from India, Pakistan, China, Turkey, Greece and Hungary. The thick enamel on the molar crown surfaces of *Sivapithecus* is in common with *Ramapithecus*, *Australopithecus* and *Gigantopithecus*; this suggests either parallel adaptations to hard object feeding or a shared ancestry separate from all other anthropomorphs (besides other tooth similarities). There seems to be a continuous gradation from *Sivapithecus sivalensis* through *Sivapithecus indicus* to the Turkish *Sivapithecus meteai*. Simons (1976) refers to *Sivapithecus* as 'ground apes' as they are generally associated with assemblages of open-country or mixed habitat, rather than the forest faunas found with the African and West European Dryopithecines. *Sivapithecus* is the most conservative of the three generas, retaining many features seen in the *Dryopithecus* species, especially in the early *Sivapithecus darwini*, whose habitats seem to have been most forested.

Sivapithecus thought to be the direct ancestor of and is closely related to *Ramapithecus*, and fossils of the two primates have often been recovered from the same deposits in the Siwalik Hills of Northern Pakistan. Other *Sivapithecus* remains have been found at sites in Turkey, Pakistan, China, Greece, and Kenya. Some authorities maintain that *Sivapithecus* and *Ramapithecus* are in fact the same species. Though *Sivapithecus* was slightly larger than *Ramapithecus*, it was only a small-to-medium-sized ape about the size of a modern chimpanzee. The fossil remains of *Sivapithecus* reveal that it shared many of the same specialized facial features of the orangutan, i.e., eyes set narrowly apart, a concave face, a smooth nasal floor, large zygomatic bones, and enlarged central incisors.

Sivapithecus's place in primate evolution was poorly understood until the 1980s. Prior to this, the genus, along with *Ramapithecus*, was interpreted as having both apelike and humanlike features and thus was presumed to be a possible first step in the evolutionary divergence of humans from the common hominoid stock of the apes. But new *Sivapithecus* finds and the reinterpretation of existing remains convinced most authorities in the 1980s that *Sivapithecus* was the ancestor of modern orangutan and diverged form the common lineage of the African apes (i.e., chimpanzees and gorillas) and humans more than 13 million years ago. The earliest *Sivapithecus* remains found so far are about 17 million years old, and the most recent are about 8 million years old.

The relationship of the modern apes of Africa and East Asia to this radiation of Miocene hominoids is of prime interest, but it is complicated. It is now fairly well established that the modern orangutan is derived from Late Miocene *Sivapithecus,* based on dental and facial similarities (Fig. 2.13). The date of the separation of this lineage from that which led to the African apes and to man is uncertain but seems likely to be earlier than 13 million years ago. The ancestry of the modern great apes is poorly known other than to say that the *Proconsul* group radiated early but seems too primitive in a number of respects to be linked directly with modern forms. The divergence of the modern apes may have occurred from 6 to 10 million years ago. The evidence obtained by molecular biologists seems to concur reasonably well on both the date of divergence of the orangutan line and that of the African great apes.

The early hominids seem certain to have originated from a Miocene ape, but it is unclear which one it might be of these that is known. It may very well be that the ancestor that is sought has not yet been discovered. The earlier view held that *Ramapithecus* was the best candidate, based on dental and palatal evidence, but such specific characteristics as thick enamel, broad, low-crowned molars, and robust jaws are shared not only with later hominids but also with several of the Middle and Late Miocene apes.

Gigantopithecus

Gigantopithecus remains from the Siwaliks of India have been dated as ca. 6.3 mya. Hominid features of *Gigantopithecus bilaspurensis* (Fig. 2.14) from the Indian Siwaliks (for instance,

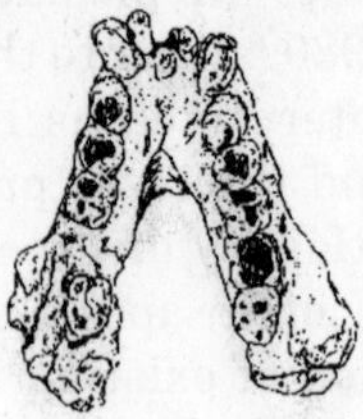

Fig. 2.14. *Gigantopithecus* Mandible

marked reduction of the front teeth, relatively small canine and tooth wear) suggest that hominid-like tendencies were already underway some 5 to 10 mya in the Pliocene period. Being considerably older than the Chinese *Gigantopithecus* (Fig. 2.15), these remains provide newer insights into the initial stages of differentiation of hominoids and man-like primates. These Siwalik finds were found associated with antelopes and primitive elephants further indicating that they inhabited open woodland areas. It can also be argued that the hominoids were displaced from India and after the emergence of Man, they made a re-entry? The Siwalik hominoids probably used ad-hoc tools to compensate for the reduction of the anterior dentition, exposure to forest ecology accompanied by dietary change and emergence of incipient bipedality (Simons and Pilbeam, 1965).

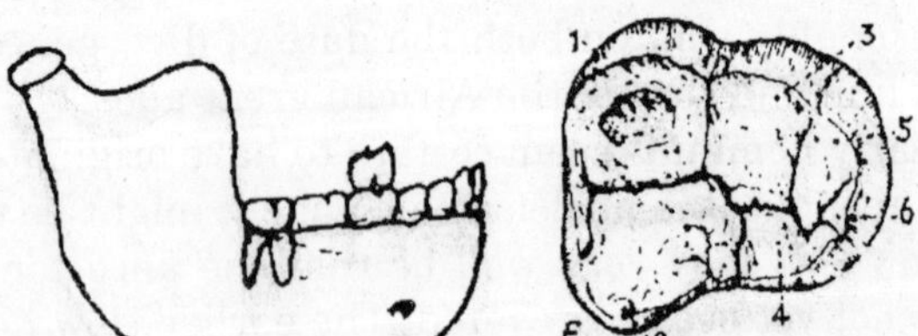

Fig. 2.15. Lower jaw of *Gigantopithecus blacki* (Left) and Lower Molar of *Gigantopithecus* (Right)

The Longgupo Cave site discovered in 1984 yielded fossil remains of *Gigantopithecus* and *Homo* as well. The former has also been reported from Liucheng Cave, Guangxi, China which is dated as greater than 1.2 mya based on amino acid racemization. The Longgupo Caves not only mark the northernmost extent of *Gigantopithecus* in China (Fig. 2.15) but also represent the third east Asian Cave where *Gigantopithecus* and *Homo* lie within the same stratigraphic interval as that of the hominid fossils (Huang *et al.,* 1995): the other two sites being Tham Khuyen Cave, Lang Son, Vietnam dated 0.475 mya and at Jianshi Cave, Hubei, China where the co-occurrence remains undated. The Longgupo site indicates the earliest presently known association; *Gigantopithecus* and *Homo* co-existed for more than 1 my in East Asia. Longgupo remains closely resemble East African fossils (*Homo habilis* and *Homo ergaster*) representing the earliest species of the genus *Homo* and share few characters in common with the Asian *erectus*. It is also contended that these Longgupo remains (0.75±0.09 my) may represent not *Homo erectus* but a more primitive species akin to *Homo ergaster,* thus far known only from Africa.

Lufengpithecus

Lufengpithecus was about the same size as a living chimpanzee, fairly generalized in structure and in many ways plesiomorphic (Etler, 1984; Kelley and Etler, 1989). It had relatively thickly enamelled cheek teeth that tend to mimic those of early hominids (Wu and Oxnard, 1983a). *Lufengpithecus* (Fig. 2.16) has been characterized as a highly sexually dimorphic

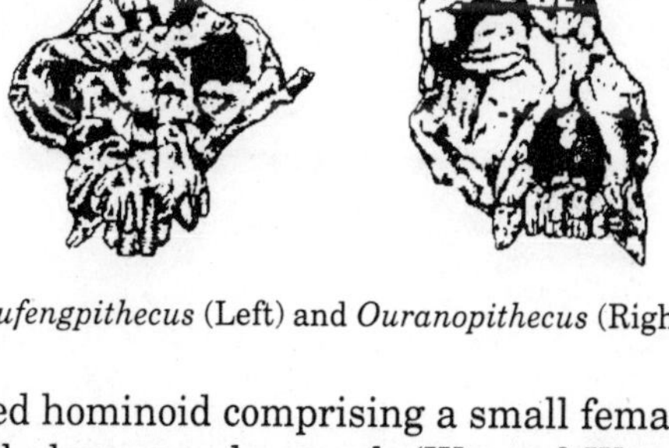

Fig. 2.16. *Lufengpithecus* (Left) and *Ouranopithecus* (Right): Frontal Views

large-bodied hominoid comprising a small female morph and a considerably larger male morph (Wu and Wang, 1987; Kelley and Etler, 1989; Kelley Xu, 1991; Wood and Xu, 1991). Its phylogenetic affinities have been controversial (see Wu and

Wang, 1988; Kelley and Etler, 1989; Schwartz, 1990) but the consensus among most workers is that it was an independent pongid lineage distinct to one degree or another from the lineage directly ancestral to *Pongo*). The small Lufeng female, however, has a number of dental traits that closely approximate those of early hominids, which led some workers to initially attribute smaller sized hominoid material from Lufeng to *Ramapithecus*, then still seen as a direct human ancestor (Wu and Oxford, 1983a, 1983b). The hominid-like dental features of female *Lufengpithecus* include relatively low-crowned canines; incipiently bicuspid P3's with tall protoconids and low-lying metaconids; a fore-shortened, lozenge-shaped P4 with protoconid and metaconid of approximately equal size; and low-crowned, thickly enamelled molars.

3

Australopithecines

Australopithecines were first discovered at the limestone quarry of Taung in South Africa. In 1924, a student of *Raymond Dart* (who was then Professor of Anatomy at the University of Witwaterstand), spotted an ape-like skull on the mantelpiece of the quarry foreman and brought it to Dart's attention. The skull was of a child aged about 3.3 years. Although Dart recognized that the skull represented an early human ancestor, his publication was met by skepticism and derision and was not taken seriously until Robert Broom found further specimens 10 years later at Sterkfontein.

Fig. 3.1. *Australopithecus* (Left-female, Right-male)

The australopithecines are only known from Africa. None has ever been found in Europe or Asia. They had ape-sized brains; their average cranial capacity ranged from 413 to 530 cc. They had strong jaws with large teeth. Like modern gorillas, the adult males were much larger than the females (Fig. 3.1). The babies probably took approximately the same length of time to grow up as a modern chimpanzee and gorilla.

Though their legs were rather short relative to their bodies, the structure of the legs and feet show that the australopithecines walked bipedally (on two legs) (Fig. 3.1) as we do, although probably not as well as modern humans. The eighty foot long track way of footprints, 3.5 million years old, discovered by Mary Leakey at Laetoli provides the most dramatic and convincing evidence of the australopithecine bipedal gait.

Australopithecus

The best known fossils from the late Pliocene and early Pleistocene (3 to 5 million years old) are those of *Australopithecus* ("Southern ape"), genus of extinct hominids, from numerous sites in Southern and Eastern Africa. Some still believe that these 'apes' were the direct ancestors of man while others regard him to represent an extinct side branch. Fossil remains suggest that australopithecines walked on two legs.

The australopithecines were distinguished from early apes by their upright posture and bipedal gait. Their brains were relatively small, not very different from those of living apes, but their teeth were more similar to those of humans. They made simple tools for cutting up of carcasses of hunted animals. These fossils are extremely important as they represent an evolutionary stage at which an ape is virtually seen merging into Man.

The Pleistocene of South Africa is rich with several man-like apes. The australopithecine skull has characteristics specific to both pongids and hominids. The hominid jaws, teeth, and palatal contour are set against a small pongid cranium with marked prognathism. Its pelvis and vertebral column are much more like those of a true man than any ape. The australopithecines are dubbed as "nut crackers" because of the large squarish teeth present in their upper jaw. They seem to have been the first hominids.

In the year 1924, Raymond Dart first reported the

discovery of these 'man-like apes' from Taung. The material comprised of a skull of a male child with full set of milk teeth and the permanent molars about to erupt. This indicates that the age of this individual was between three to four years and had a cranial capacity of around 405 cc. It took 73 days to retrieve this skull from the limestone deposits at Taung. Dart (1925) described this material as not having a very human face but instead it had protruding jaws and therefore known as a

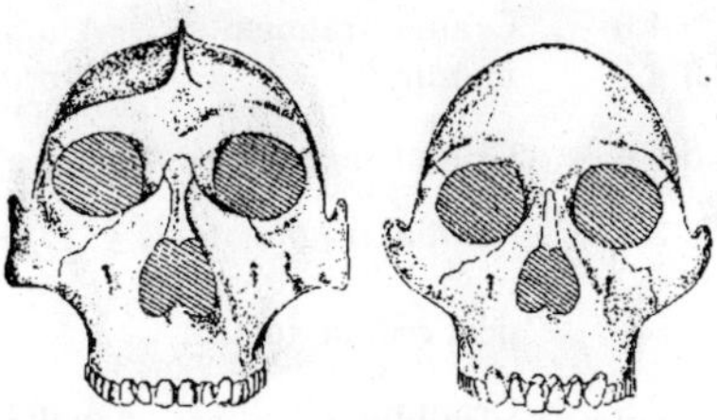

Fig. 3.2. Reconstruction of *Australopithecus* skulls; left-Sterkfontein; right-Swartkrans

'man-like ape'. He further argued that these fossils represented an extinct race of apes intermediate between living anthropoids and man. He designated these fossils as that of *Australopithecus africanus*. These fossils remains of *Australopithecus* possess the essential characteristics of genus *Homo*: a largely upright posture, bicuspid molars and reduced canines (in size). The South African sites of (Table 3.1) of Sterkfontein, Swartkrans, Kromdraai and Makapansgat yielded remains of skulls (Fig. 3.2), jaws, many teeth (Fig. 3.3) and limb bone fragments. These finds mostly represent two forms: a smaller form (gracile) and a larger type (robust).

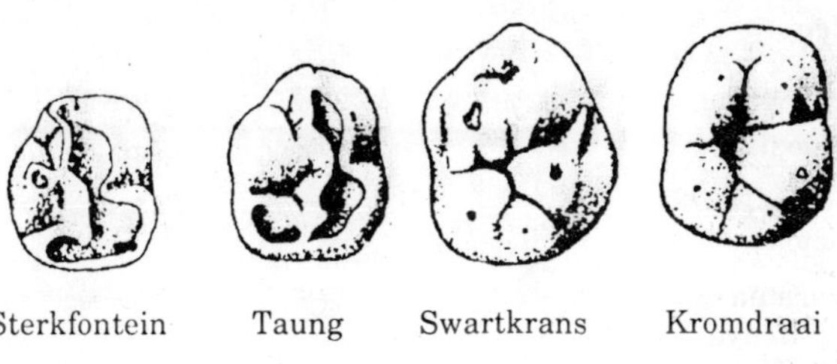

Fig. 3.3. Deciduous Dentition of *Australopithecus* (First lower molars in increasing order of molarization)

TABLE 3.1
Some Important Early Hominid Sites

Site	*Age (mya)*	*Fossils*	*Species*
Lothagam, Kenya	5.5	Mandible fragment	*Australopithecus* sp.
Tabarin, Kenya	4.5	Mandible fragment	*Australopithecus* sp.
Kanapoi, Kenya	4.0	Humerus	*Australopithecus* spp.
Middle Awash, Ethiopia Belohdelie	4.0	Cranial fragments	*A. afarensis*
Maka	3.5-4.0	Femur	*A. afarensis*
Hadar, Ethiopia	2.8-3.4	Crania, mandibles	
Omo, Ethiopia	1.8-3.1	Cranial fragments, mandibles, post crania, teeth	
Lomekwi, Kenya (West Turkana)	2.5	Cranium	*A. aethiopicus*/ *A. boisei*
Other West Turkana sites	1.8-3.4	Crania, mandibles post crania	
Laetoli; Tanzania	3.6	Cranial fragments, mandibles, post crania	*A. afarensis*
Makapansgat, South Africa	2.5-3.0	Crania, mandibles post crania	*A. africanus*
Sterkfontein South Africa	2.5-3.0	Crania, mandibles Post crania	*A. africanus*
Taung, South Africa	? 2.0+	Cranium	*A. africanus*
Swartkrans, South Africa	?1.5-1.8	Crania, mandibles, post crania	*A. robustus*
Kromdraai, South Africa	?2.0	Cranium, mandible, post cranial	*A. robustus*
East Africa			
Garusi, Tanzania	-	-	*A. afarensis*
Hadar, Ethiopia	-	-	*A. africanus*
Olduvai Gorge, Tanzania Omo, Ethiopia	-	-	*Paranthropus*
Peninj, Tanzania Koobi Fora, Kenya Chesowanja, Kenya	-	-	*A. boisei*

Both, the gracile and the robust australopithecines had a small brain and appeared on the scene around 5.6 million years ago. This South African fossil material came from three main sources (limestone dumps/ hills due to quarry operations from rock piles, fissures, and cave deposits) scattered in three widely separated regions:

- Taung – a dolomite plateau (limestone/marble) having deep cracks into which animal bones had fallen and got cemented into two layers: grey below and pink above.
- Sterkfontein – a cave site with a hole in the roof.
- Three other sites – ordinary caves, i.e., without a hole in the roof. None of these five sites can be designated as places of habitation or occupation.

The East African fossil material, mainly from Ethiopia, Tanzania and Kenya, represent three varieties of the 'man-like apes':

- small brain, short statured, bipedal of female 'LUCY' from Hadar near the Awash river in Northern Ethiopia;
- this material is relatively primitive and dated as 3.5 million years ago;
- larger type (robust australopithecine), and
- smaller form (gracile australopithecine).

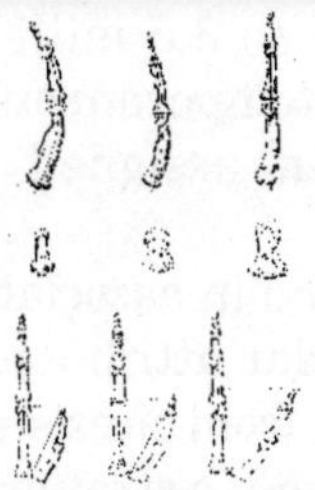

Fig. 3.4. Bones of the hand : Left – Juvenile Gorilla, Middle – Olduvai Hominid, Right – Modern Man

Stone tools were found associated with one of the skull fragments from Koobi Fora in Kenya. However, the tool industry dates to about 2.5 million years ago after the first human remains were reported from Hadar. Fossil material from Hadar, probably of a family of thirteen individuals, consists of jaws, teeth, leg bones, hand (Fig. 3.4) and foot bones

(Fig. 3.5), partial adult skull and an infant's nearly complete lower jaw. In the year 1981, forty-five miles south of Hadar, five skull fragments and an upper part of the femur were discovered. These remains are older than those found at Hadar. The femur suggests that the individual took to bipedalism. The skull, on reconstruction, seems to have a cranial capacity of about 400 cc.

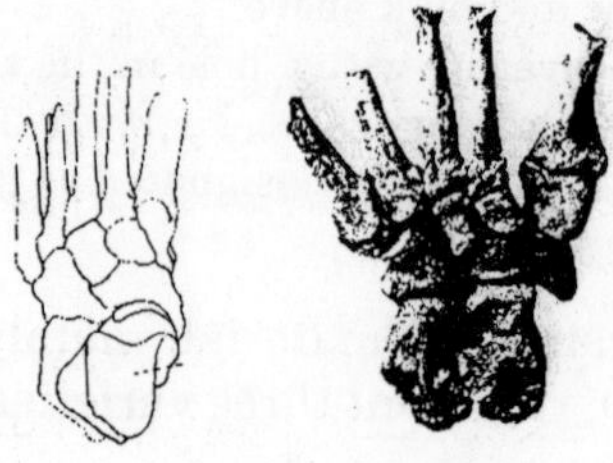

Fig. 3.5. Left foot bones (upper view); *Australopithecus* (left) and *Homo habilis* (right); foot bones of *Homo habilis* suggest upright walking

A rich collection of hominid teeth and jaw fragments, of about thirteen individuals, mostly surface finds, was made from Laetoli, around thirty miles south of Olduvai Gorge. This is a wooded site (3.77 to 3.5 million years ago) near a volcanic fountain and is contemporaneous with that at Hadar. Laetoli remains have been assigned to *Australopithecus afarensis*.

No tools have been found in association with the hominid fossils from Laetoli and Hadar attributed to *Australopithecus afarensis*. The earliest tools, from sites such as Koobi Fora and Olduvai Gorge, have been found in stratigraphic levels that have yielded robust australopithecines, *Homo habilis*, and – in the case of Koobi Fora – *Homo erectus*. Under these circumstances, it is not possible to decide which of the hominid species was the first stone toolmaker.

The fossil material from Africa represents such a wide variety of australopithecine forms that they could be classified into separate genera and species:

- *Australopithecus africanus* – a lower Pleistocene gracile form from South Africa.

- *Australopithecus boisei* – a lower Pleistocene hyper-robust form from East Africa.
- *Paranthropus* – a predominantly middle Pleistocene robust variety from South Africa.

A partial jaw found in Chad (Central Africa) greatly extends the geographical range in which australopithecines are known to have lived. The specimen, who has been nicknamed Abel, has since been named *Australopithecus bahrelghazali.* (Brunet *et al.,* 1995).

The earliest and most primitive of these is the *Australopithecus africanus* (5.5 to 2 million years ago) (Fig. 3.6). The larger and the more evolved robust forms came later: *Australopithecus robustus* (South Africa) and *Australopithecus boisei* (East Africa) (Fig. 3.7). They were on the scene from 2.5 to 1 million years ago. Another gracile form, the *Australopithecus afarensis*, popularly known as 'Lucy' is dated as 3.5 million years old. The genus, *Paranthropus*, an aberrant hominid, arrived on the evolutionary scene when tool making was already an established industry. This latter form is a widely spread and highly specialized type of the hominid.

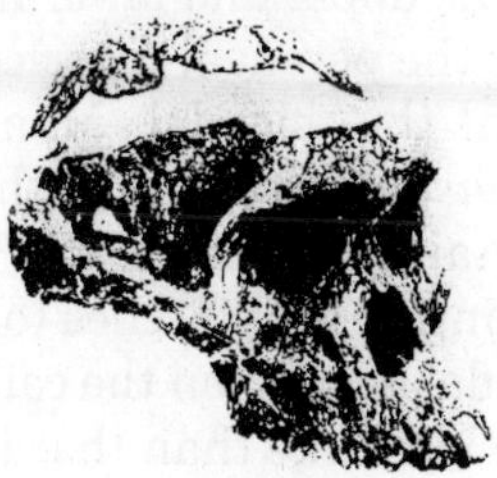

Fig. 3.6. Skull of *Australopithecus africanus*

Size of the knee joint in the australopithecines compared to the sizes of the various joints of their upper limbs implies a sharing of the body weight bearing between them (Oxnard, 1975). Johansson (1980) regards Lucy's knee to be essentially human in its form suggesting thereby human-like bipedality in their locomotion (Fig. 3.7). The extent to which the leg is placed laterally at the knee (valgus form of the knee) is an important indicator of bipedality because it enables the foot to be placed more directly under the center of gravity of the body; this posture provides better balance and more time for the free leg to swing forward during walking. Tardieu (1981) believes that the form of the knee joint did not allow the leg to swing as far forwards as that of humans suggesting that their stride was short.

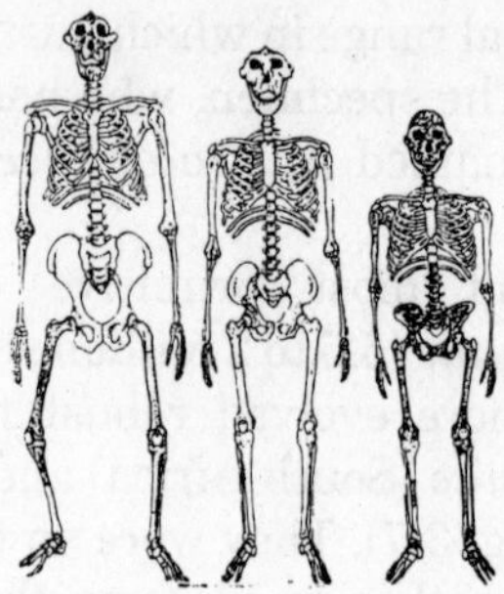

Fig. 3.7. Australopithecines; left-*A. boisei*; Middle-*A. robustus*; right-*A. africanus / habilis*

The relative sizes of the lower leg joints in the australopithecines were more compatible with sharing of weight bearing between both the upper and lower limbs (Fig. 3.7). The groove on the fibula for the peroneal muscles (which help evert the foot) is small in the bipedal Man but is large in *Australopithecus afarensis* (Lucy) as in the other apes; this enables Lucy to maintain an arched foot. Thus, this is the evidence of the beginning of bony arches (on the plantar side) for supporting the bipedal gait. Even the calcaneum or the heel bone is large and more man-like than that in the apes.

Four australopithecine foot bones dated at around 3.5 million years are the oldest hominid fossils yet found in South Africa. They seem to be adapted to bipedalism, but have an intriguing mixture of ape and human features (Clarke and Tobias, 1995). Since then, eight more foot and leg bones have been found from the same individual, who has been nicknamed Little Foot.

Analysis of new *Australopithecus africanus* fossils from Sterkfontein in South Africa suggests that the forelimb and hind limb proportion of *africanus* were more ape-like than in the earlier *Australopithecus afarensis* (McHenry and Berger, 1998).

A fragmentary skull of a new species, *Australopithecus garhi*, found by a joint Ethiopian, American and Japanese team near Bouri, an Ethiopian village in the Middle Awash region, northeast of Addis Ababa, could well be from the missing australopithecine that sired the human race. Excavated in 1997, its jutting face and upper jaw filled with large teeth clearly belongs to a species more advanced than *Australopithecus afarensis* (Fig. 3.8) yet more primitive than the earliest humans. This small-brained, large-toothed hominid was found near antelope bones, which had been butchered by stone tools (Asfaw *et al.*, 1999).

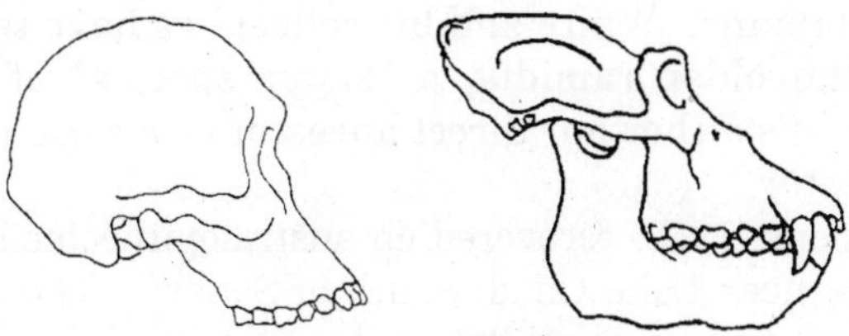

Fig. 3.8. Skulls of *Australopithecus afarensis* (left) and Gorilla (right)

The mix of characters wasn't precisely what they were looking to see: smaller, more specialized teeth and a larger braincase. So, they named their hominid *Australopithecus garhi* (*garhi* means surprise in Afar). But the skull's intermediate anatomy and its age – about 2.5 million years – put it midway in both time and form between the most recent *Australopithecus afarensis* (Fig. 3.8) and the oldest known fossils of our own genus.

Then, in 1994 and 1995, teams working in Ethiopia and Kenya announced that they had each found a new species of hominid. Both discoveries smashed the 4 million-year barriers. An international team in the Middle Awash region of Ethiopia dug up the first – and at 4.4 my, the oldest, about 50 miles south of where Lucy was discovered.

All told, the scientists excavated the bones and teeth of 17 individuals. Given their age, no one was surprised that they showed a mix of chimpanzee-like and human traits that as a whole are more primitive than those of *Australopithecus afarensis*: smaller molars, larger canines and thinner tooth enamel, suggesting a diet rich in easy-to-chew fruits and vegetables. The new species, says palaeontologist Tim White of the University of California at Berkeley, a co-leader of the expedition, "is way closer to an ape than to an australopithecine and is significantly different from any other hominid."

Because the fossils were too distinctive to be included in Lucy's extended *Australopithecus* family, the researchers called the new species *Ardipithecus ramidus* (*ardi* means ground or floor in the local Afar language, and *ramid* means root). Precisely where do *Australopithecus ramidus* and *Australopithecus anamensis* fit into the scheme of human evolution? Leakey believes the latter is a direct ancestor of *Australopithecus afarensis* and thus a direct ancestor of modern humans. White and his colleagues have tentatively labeled the older ramidus a "sister species" of all later hominids; it's either our direct ancestor or a close relative of that ancestor.

Coppens (1962) recovered an australopithecine skull from Koro Toro, near Lake Chad, South of Sahara. The associated fauna suggests an early Villafranchian age, i.e., no pebble tools are associated with the fossils although pebble tools are present in the later Villafranchian deposits of North Africa). This skull find is stated to be intermediate in morphology between *Australopithecus africanus* and *Australopithecus robustus* but perhaps has a greater cranial capacity (Arambourg, 1963). Robinson (1963) regards the Koro Toro skull to be closer to *Australopithecus robustus.*

Leakey and co-workers (1959, 1961, and 1964) described a number of hominid finds from Olduvai Gorge in Tanganyika belonging to early and middle Pleistocene. This material has been assigned to two distinct species:

- *Australopithecus boisei* (*Zinzanthropus*) from Bed I at Olduvai and from middle Pleistocene deposits near in Tanganyika
- *Homo habilis* from Beds I and II at Olduvai (Dart calls him "handy man").

Australopithecus boisei is more bigger and robust than *Homo habilis* (Fig. 3.9) and has larger premolars and molars. Simon and Pilbeam (1972) believe that *Homo habilis* cannot be distinguished at the specific level (Bed I) from *Australopithecus robustus* (Fig. 3.9). The *Homo habilis* material falls into two groups, Bed I and Bed II, separated by as much as one million years. Again, during the period of deposition in the lower parts of Bed II, two distinct taxa appear to have been present: *Homo habilis* at Olduvai and *Australopithecus boisei* at Lake Natron. A new *Australopithecus boisei* skull is one of the most complete known, and the first known with an associated cranium and lower jaw. It also has a surprising amount of variability from other *boisei* skulls, which may have implications for how hominid fossils are classified (Suwa *et al.,* 1997; Delson, 1997).

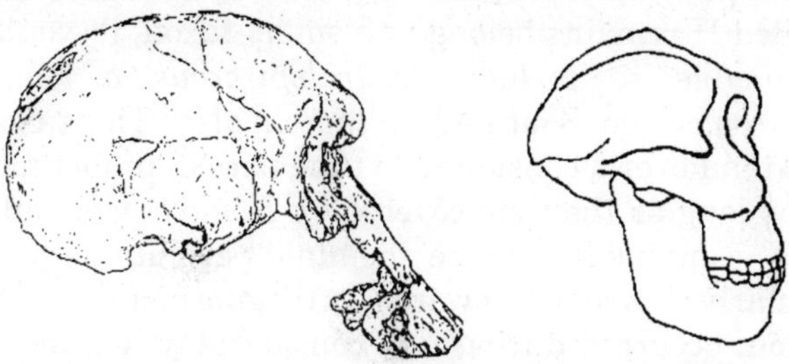

Fig. 3.9. *Homo habilis* KNM ER 1470 (left) and *Australopithecus robustus* (right)

Half a dozen supposedly distinct taxa, based on most fragmentary and limited material, have been proposed for the early and early-middle Pleistocene hominids. *Australopithecus robustus* (Fig. 3.9) is known from South Africa (Swartkrans and Kromdraai), Olduvai Gorge bed I, Lake Natron and possibly Java. Other African forms, of roughly equivalent age, have been referred to three taxa: *Australopithecus africanus, Telanthropus capensis* and *Homo habilis*. The relative dating of North, East and South African sites presents number of problems. The South African sites which have yielded *Australopithecus africanus* remains are broadly contemporaneous with the later Bed I levels at Olduvai Gorge (Cooke, 1963). *Telanthropus capensis* from Swartkrans is probably contemporaneous with Bed II *Homo habilis.* Probably these two, i.e., *Telanthropus capensis* and *Homo habilis*, represent two populations of a single species/subspecies. However, more material is required for firm conclusions to be made. Both are, however, similar to the mandible of *Homo erectus* from the Djetis beds in Java (so called '*Pithecanthropus* B').

Primitive stone tools of uniform type associated with *Homo habilis* throughout Bed I and the lower strata of Bed II correlate with the apparent stability of the cranial capacity in these strata fossils. However, Napier (1962) suggests that the hand of *Homo habilis* was not as refined, structurally and functionally, as that of the later men.

Australopithecus africanus and *Homo habilis* from Bed I are not specifically distinct. Most of the Bed II fossils and *Telanthropus* might be specifically identical. If the Olduvai Bed I and Bed II remains belong to a single taxon, it would not be unreasonable to include *Australopithecus africanus* and *Telanthropus* from South Africa therein also. This taxon would then extend over a considerable time period, in fact more than twice as long as the time covered by *Homo erectus* and *Homo sapiens* combined. Some hominid evolutionary change, particularly tendency towards orthognathism and reduced dentition, occurred during this considerably lengthened time period, although the extent of variation within the taxon was probably not greater than that observed in modern mammalian species including man.

Australopithecus robustus and *Homo habilis* were evidently coexistent in the same general areas for perhaps a million years, if the new dates and the stratigraphic faunal data of Leakey are right. They were, therefore, sympatric species. Both were bipeds, presumably tool users and toolmakers. Comparative study of other closely related pairs of animal species indicates that *Australopithecus robustus* and *Homo habilis* must have had separate origins in different geographic regions, i.e., originally they must have been allopatric (Kohn and Orians, 1962). Probably, one hominid species evolved in Africa and the other in Asia and mingled at the beginning of the Pleistocene epoch when land connections between Eurasia and Africa presumably became established. Tobias (1965) believes that 'at least in East Africa and probably, too, in South Africa, *Homo habilis* and *Australopithecus* spp. were sympatric and synchronic'. He further concludes that 'since they are contemporary by actual fossils recovered to date are clearly too late – and possibly slightly too specialised – to have been in the actual human ancestral line unless we are to postulate a polyphyletic origin of Homininae at varying times from the australopithecine stock'. Tobias, however, believes that the 'closeness of morphology between *Homo habilis* and *Australopithecus africanus* points strongly to a common ancestry in the upper Pliocene or the very beginning of Pleistocene'.

Dental analysis reveals that the molars of *Homo habilis* are similar in size and shape with those of *Australopithecus africanus* from Sterkfontein. The premolars of the former are somewhat narrower than those of the South African forms (Robinson, 1956). Many of the Olduvai specimens are crushed and broken and thus limits their comparisons. *Homo habilis* and *Australopithecus africanus* perhaps represent two variant populations within the same widespread species but this can only be corroborated when more evidences are available.

Validity of the *Homo habilis* depends on the reliability and plausibility of its diagnosis/identification/assessment. Campbell (1964) states that 'it is here that the hypothesis of the new species must stand or fall; the diagnosis must support the hypothesis for the species to stand, not in law, but in reality . . . . (In their original publication, the authors stated that *Telanthropus capensis* 'may well prove to be, on closer examination, to belong to *Homo habilis*'). *Telanthropus capensis* from Swartkrans is probably contemporaneous with Bed II *Homo habilis*, broadly speaking.

But their (*Australopithecus's*) nasal morphology is very similar to that in the anthropoid apes: lack nasal bone convexity, i.e., have a flat, non-protruding nasal skeleton. The australopithecines have a relative large brain size as in the extant apes (not advanced much beyond them), and are dependent on large post canine teeth for processing food; mixed climbing with bipedalism.

Early Species

By the 1980s, it was postulated from fossil evidence that the evolutionary divergence of australopithecines (as these early hominids are often collectively known) from apes occurred about 8 million years ago. By about 1990, comparative molecular studies of DNA from the cells of present-day humans and chimpanzees suggested that the two groups diverged from a common ancestor about 6 to 4.5 million years ago. Fossil records of hominid remains earlier than about 3.75 million years ago were unknown, however, until 1994, when the discovery of a previously unknown species of hominid was announced. Originally, designated *Australopithecus ramidus*, the species was later reclassified under a new genus as *Ardipithecus ramidus*. Its fossil bone fragments had been found at ARAMs, in north central Ethiopia, in 1992-93. These remains, which dated from about 4.4 million years ago, provided indirect evidence of an upright stance and of bipedal walking (as opposed to the knuckle walking of chimpanzees). The character of its dentition (teeth) shared affinities with both early apes and later hominid species.

Based on fossil remains uncovered in Kenya at Kanapoi (1994) and at Allia Bay (1988), *Australopithecus anamensis* was named in 1995 by Meave Leakey. Skull fragments, particularly teeth, attributed to the species are similar to those of early apes, yet the arm and leg bones are more developed and indicative of bipedalism. Its remains were dated between 3.9 and 4.2 years old.

The characteristic features of *Australopithecus africanus* (Taung child) are:

- well preserved face with its upper part flat;
- no frontal flattening; high forehead;
- less prominent brow ridges;
- prominent glabella;
- large and rounded orbits;
- flat nasal bones giving the face a 'dish-shaped' appearance, short face;
- squarish nasal aperture with no nasal spine;
- well developed premaxillae;
- subnasal prognathism of the maxilla, i.e., prognathus jaw;
- regular parabolic dental arcade; anterior teeth are relatively large;
- sloping symphyseal region, no chin, no simian shelf; diastema is present;
- cranium seems to be globular with forwardly directed foramen magnum;
- cranial capacity in between 400 to 500 cc;
- weight in between 30 to 60 kg; dated as about 3 to less than 2.5 million years old; and
- the characters seem to be very similar to those of chimpanzees and gorillas.

The Robust Australopithecines

A subgrouping within the australopithecines, the robust australopithecines are sometimes referred to the genus *Paranthropus* rather than *Australopithecus*, because they are so readily distinguished from the more lightly built species of *Australopithecus*. Robert Broom announced the discovery of the first robust australopithecine in 1939 from Kromdraai in South Africa. Three species of robust australopithecines are recognised: *Australopithecus robustus*, which is confined to Southern Africa, and *Australopithecus aethiopicus* and *Australopithecus boisei*, both confined to East Africa and Ethiopia.

The majority of the *Australopithecus robustus* remains have been found at the South African sites: Swartkrans and Kromdraai and, dating from just over 2 million to 1.5 million years.

In East Africa, the earliest robust australopithecine, *Australopithecus aethiopicus* is found at sites between 2.6 and 2.3 million years old west of Lake Turkana and in the Omo Valley. *Australopithecus aethiopicus* was most probably the ancestor of *Australopithecus boisei*.

Australopithecus robustus has the following distinguishing features:

- heavily built skull with a sagittal crest present sometimes;
- higher skull vault and rounded appearance; vertex is relatively much higher compared to the face than in the apes;
- long, broad, quite heavy and massively constructed face; especially the jaws, e.g., much forward than in the living humans but does not project as in the pongids; moderate facial buttressing; sloping forehead;
- massive brow ridges;
- nuchal ridges are low compared to the apes;
- anteriorly placed foramen magnum and the occipital condyles; as in genus *Homo*;
- arcuate tooth row (rounded dental arcade) with no diastema; no simian shelf; very thick jaws; small and spatulate incisors and canines; large premolars and very large molars; hominid-like;
- well developed mastoids;
- cranial capacity estimated as 530 cc;
- lumbar curve; short and broad ilium; iliac crest has a sigmoid curve; deep sciatic notch; short pubic symphysis; short and broad

sacrum; short sacro-iliac joint;
- basin shaped pelvis;
- hominid like talus; plantigrade (heel and sole squarely on the ground); these characters suggest that *Australopithecus robustus* was an orthograde bipedal hominid;
- weight in between 40 to 80 kg;
- height ranging from 1.1 to 1.3 metres;
- moderate sexual dimorphism; and
- existed two to one million years ago.

Australopithecus boisei is characterised by features, which are diagnostic of the family Hominidae, namely:

- indications of cerebellar enlargement;
- appreciable flexion and anterio-posterior reduction of the cranial base – relatively orthognathus face, low spheno-occipital index, markedly elevated positioning of the porion;
- relatively anterior displacement of the occipital condyles and the foramen magnum;
- nearly horizontal plane of the foramen magnum;
- vertical slope of the posterior face of the petrous pyramid;
- detailed structure of the external acoustic porous and the meatus, tympanic plate and the supramastoid crest;
- structure of dorsum sellae and sella turcica of the sphenoid bone;
- orbits of only moderate height, despite the marked overall tendency towards facial elongation;
- presence of elevated anterior nasal spine;
- reduction of the subnasal prognathism; and
- lateral position of the mandibular fossa indicating that, despite condylar, dental and mandibular enlargement, the space in between the two halves of the mandible has not been appreciably reduced.

The Gracile Australopithecines

The gracile australopithecines are the earliest known true hominids. Recently, a new species, which was initially believed to be an australopithecine and called *Australopithecus ramidus*, was found at a 4.4 million year old site called Aramis in the Awash Valley in Ethiopia. Later, Tim White and Berhane Asfaw who led the expedition that discovered this species, recognised that it was a new genus and gave it the generic name *Ardipithecus*. This species may prove to be the first true hominid, but it has not yet been fully studied.

There are several species of *Australopithecus*, ranging in

age from 4.1 to about 1.0 million years old. The earliest australopithecine, *Australopithecus anamensis*, was found at Kanapoi, a site south west of Lake Turkana, Kenya and also at Allia bay, a site to the east of the lake. *Australopithecus anamensis* lived between 4.2 and 3.9 million years ago.

Australopithecus anamensis may be ancestral to *Australopithecus afarensis* which is better known at Laetoli in Tanzania, where Mary Leaky discovered the 80 foot long trail of footprints, and at Hadar in Ethiopia where Johanson discovered the almost complete skeleton known as "Lucy" as well as more than 200 other specimens of this species.

Australopithecus afarensis lived between 3.6 and 2.9 million years ago. In South Africa *Australopithecus africanus* is known from Makapansgat and Sterkfontein, sites believed to date between 3.5 and 2.0 million years ago. The robust australopithecines, *Australopithecus boisei*, *Australopithecus aethiopicus* and *Australopithecus robustus*, lived in East and South Africa between 2.5 and about 1.0 million years.

Paranthropus

A genus of the family Hominidae, *Paranthropus* first occurred in Omo succession at least 2.1 million years ago (Howell, 1982) but does not appear in South Africa until about 1.5 million years ago (Brain, 1982). However, they were present at an earlier date in East Africa than in the South. They are a widely spread and a highly specialised form of the hominid, had brain larger than that of *Australopithecus* yet not as large as that of *Homo*. *Paranthropus* (Fig. 3.10) is characterised by the formation of a slightly concave and low forehead with a frontal trigone delimited laterally by posteriorly converging temporal crests, a central facial hollow with a completely flat nasal skeleton, cheek region situated anterior to the plane of the pyriform aperture, and the nasal-alveolar clivus sloping smoothly into the nasal cavity floor. The temporal fossa is capacious and mediolaterally expanded. Small incisive canals open into the horizontal nasal floor, premolars are greatly enlarged relative to the canines and molars, molars greatly enlarged, the anterior teeth being smaller compared to the post canine dentition, low and bulbous cusps of the cheek teeth and a massive tooth bearing bone are characteristic features of

Paranthropus. Their palate tends to be shallow anteriorly and deep posteriorly, in these specialized forms of the hominids, glabella is situated at a lower level than the supraorbital margin and have a flattened "rib" of the bone across each supraorbital margin.

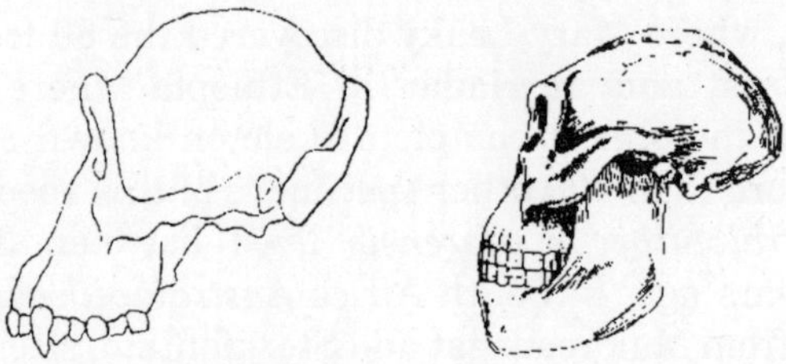

Fig. 3.10. Comparison of skulls of a modern ape (Left) and *Paranthropus* (Right)

Many people are now using the genus name *Paranthropus*, originally given to *robustus,* to refer to the robust australopithecines (*robustus, boisei*, and a*ethiopicus*). This change makes sensc if all these species form a clade (all of the species descended from a common ancestor) but it is not yet known if this is the case. *Australopithecus* is an immediate evolutionary precursor of the "*Pithecanthropus*" phase of the hominid evolution.

4

Homo habilis

The first *Homo habilis* (handyman) fossils (Fig. 4.1) were found in 1959-1960. *Homo habilis* lived nearly 2 million years ago. This was determined by potassium-argon dating which tests the amount of argon in relation to radioactive potassium in volcanic rocks to minerals. These fossils were found in Olduvai Gorge in Tanzania, Africa. This species came before *Homo erectus* and after *Australopithecus* in the fossil record.

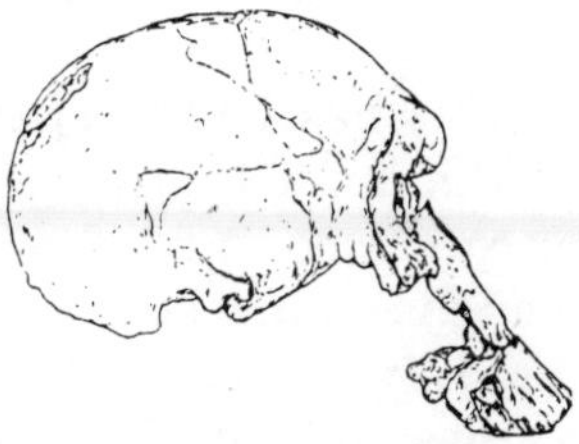

Fig. 4.1. *Homo habilis* KNM ER 1470

In the year 1972, a dramatic new discovery was made on the shores of Lake Rudolf in north Kenya. The skull (Fig. 4.1) and leg bones of a creature provisionally named '1470 Man' (National Museum of Man, Kenya Catalogue number), suggests that another tool maker, with a large brain and more human like features than *Australopithecus*, lived nearly 3 million years ago. '1470 Man' is probably the man's earliest ancestor. Habilines are classified as human and given the Latin name *Homo habilis*.

Recovery of fossils of a more gracile form, of more than one individual, from Bed I of Olduvai Gorge associated with stone tools and evidence of some social organisation around a 'homo base' display a mosaic of primitive and advanced traits suggesting that it walked erect and bipedally. Leakey (1961) assigned this material to a new hominid taxon, *Homo habilis*. This suggests that it was the true precursor of modern *Homo* and coexisted with the genus *Australopithecus*. However, it is the brain size relative to the body size, which is of significance. Habilines have the greatest average brain size (approx. 600cc) relative to body size compared to the australopithecines. These remains came from about the same level 300 feet away, from a slightly lower level (2 feet) than the *Zinzanthropus* site, and seem to belong to a more advanced type of early humans. Fossils OH7, OH8 and OH 35 represent the earliest members of this lineage at Olduvai. They are dated to be 1.75 million years ago.

The hand bones are consistent with the tool making abilities. There is evidence that the hands of *Homo habilis* were capable of "fine manipulation". As a result, all tools found in east Africa are now attributed to *Homo habilis*. *Homo habilis* is believed to be a hunter from the evidence of stone tools and animal bones found. This has been disputed and it is now believed that *Homo habilis* hunted small game animals but was a scavenger of larger carcasses. It has been found that there are small cuts marks on the animal bones indicating that stone tools were in fact used to butcher these animals, but not necessarily in hunting these animals. The teeth of *Homo habilis* are considered to be most modern than those of their contemporary hominids. Even their skulls are less robust compared to the robust australopithecines and have a less prognathic face. The nasal bones, although incomplete, appear to be convex and posses a slight internasal angle (Franciscus and Trinkaus, 1988).

Two more skulls of delicate built have been found from Olduvai Gorge in Tanzania (East Africa). These together with the '*habilis*' variety resemble *Australopithecus africanus* in details of the type of the mandible and the small brain size. This suggests the presence of a third form of the hominid still unreported from the late Pliocene or early Pleistocene of East Africa. Leakey and Leakey (1978) propose that at least three hominid taxa were contemporary at Koobi Fora in Lake Turkana.

Homo habilis lived alongside the robust australopithecines in East Africa about 2 to 1 million years ago. They represent an intermediate stage between the *Australopithecus* and *Homo erectus*. *Homo habilis* retained fully the *Australopithecus* sized face and dentition. It is argued that *Homo habilis* represented a very old human evolutionary lineage and was on the direct line to modern *Homo* while the early hominids (including *Australopithecus*) were not. The contemporaneous australopithecines are believed to represent an evolutionary dead end. A minority view is that *Homo habilis* is just a gracile australopithecine and regards *habilis* as an ancestral member of the gracile line. Because of the morphological variability seen in the fossil material attributed to *Homo habilis*, there is lot of controversy about its either being a polymorphic species or more than one species. Critical analysis of the endocranial volume does not support the multiple species hypothesis but instead suggests that *Homo habilis* represents a single sexually dimorphic, and perhaps a polytypic species (Miller, 1991). The species *Homo habilis* is now generally regarded as a transitional form ancestral to all later *Homo* (Cartmill et al., 1986); it represents an intermediate stage between *Australopithecus* and *Homo erectus*; and it has retained fully the *Australopithecus* sized face and dentition.

Sites

Tanzania. Olduvai Gorge – eight specimens including several teeth, jaws and skull fragments, hand bones and a foot (1.8 to 1.75 million years ago).

Ethiopia. Omo River Valley – many isolated teeth, five jaws, two skulls, femur and a part of the ulna.

Kenya. Lake Turkana – skull and postcranial bones (1.9 to 1.8 million years ago).

The fossil *Homo habilis*, a topic of contention since its first description in 1964, a short lived African species dated around two million years ago. *Homo habilis* is the first hominid with a distinctly large brain (600 to 800 cc), and its poorly known postcranial skeleton appears to be more like that of later *Homo* than is any australopithecine's. Yet, it differs little from earlier *Australopithecus* in its face, jaw and teeth.

Peter Nzube discovered *"Twiggy"*, *Homo habilis* (OH 24) in 1968 at Olduvai Gorge in Tanzania. It consisted of a badly crushed skull and seven teeth. It is about 1.8 million years old and has a brain size of about 590 cc. This is the most complete of the *Homo habilis* skulls discovered at Olduvai Gorge. When discovered, it had been crushed absolutely flat, hence the nick-name *"Twiggy"*, after an English model of the 1960's.

Fossil remains of *Homo habilis* discovered from Bed I and Bed II of Olduvai Gorge can be grouped into two main groups (A and B), separated perhaps by a million years from one another. The type remains consist of:

- Remains from Bed I (this site is little lower than the site of Olduvai hominid 5 and is thus also known as pre-*Zinzanthropus*);
- a juvenile mandible (about 12 years in age);
- two parietal fragments;
- bones of the hand; and
- an upper molar tooth.

The foot bones and clavicle, belonging to an older individual, were also discovered from another site FLKNN I. Two other sites in Bed I have yielded teeth, mandibular and skull fragments, a tibia and a fibula belonging to either *Homo habilis* or to *Australopithecus boisei*.

Bed I has been dated as 1.75 million years before present by Potassium-Argon method and as 2.03 million years ago by the Fission Track method; the two dates being quite close to each other.

Remains from Bed II comprise of:

- a complete mandible with an associated maxilla;
- occipital, parietal fragments and temporals of possibly the same individual;
- isolated teeth; and
- a damaged palate.

Tobias (1964) suggests a cranial capacity between 642 cc to 724 cc based on a reconstruction of the crania from the two parietal fragments found from Bed I. He had earlier suggested a cranial capacity of 435 cc to 600 cc for the australopithecines in the year 1963.

The sequence through Bed I to Bed II suggests that the teeth and mandibles became progressively reduced during this time whereas there was little alteration in the cranial capacity. The mandible and the juvenile teeth from FLKNN I in Bed I are about ten percent larger in all measurements than the mandible from MNK II in Bed II.

Australopithecus africanus and *Homo habilis* both were habitual bipeds. Unfortunately, pelvis of the former but not of the latter was discovered. While the foot of the latter and not of the former had been found. It is possible that the foot (Fig. 3.5) of *Homo habilis* from the Olduvai Bed I is no more nor less *Homo*-like than is the pelvis of *Australopithecus africanus* from South Africa. Teeth and mandible of *Homo habilis* from Olduvai Bed I FLKNN do not differ greatly from those of *Australopithecus africanus* in morphology.

Homo habilis remains from Bed I and lower parts of Bed II at Olduvai, in terms of the size of their braincase and consequently of its endocranial capacity, occupies a position in between the biggest-brained australopithecine specimens available and the smallest-brained cranium of *Homo* hitherto discovered. *Homo habilis* departs from the australopithecinae fossils in the shape of its teeth, especially the premolars and lacks the characteristic bucco-lingual enlargement of their mandible. Though *Homo habilis* is associated with hominised hands and feet, it lacks the cranial tori and thickening characteristic of *Homo erectus*.

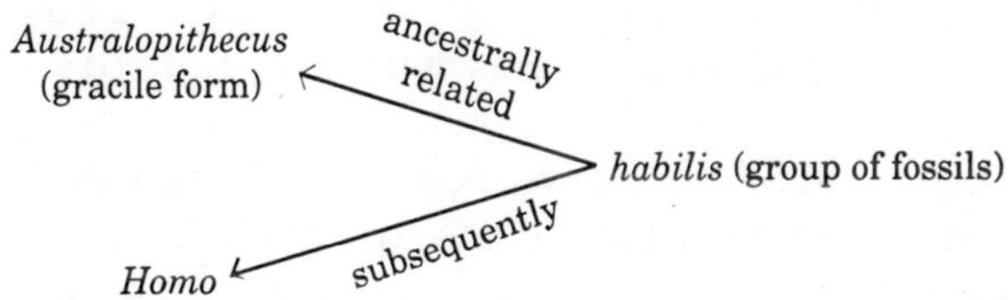

Different views have been expressed about the status of these fossil remains from Bed I and Bed II at Olduvai. Ewer (1967) suggests clubbing together of *Australopithecus* and *Homo habilis* as *Australopithecus/Homo habilis*. He assigns the material to the genus *Australopithecus* as a new species, namely *Australopithecus habilis*. Others would create a new genus between *Australopithecus* and *Homo*; and few would assign the fossils to an existing species of *Homo*, i.e. *Homo erectus* as a new subspecies, *Homo erectus habilis*. Tobias (1965) suggests that *Homo habilis* may be the most effective link between the australopithecines and *Homo ere*ctus and may, thus, close this gap. Howell (1967) found a series of isolated teeth, including ultragracile *Homo habilis* teeth, at Omo in the north of Lake Rudolf. He believes that *Homo habilis* is indeed a valid discrete taxon. Tobias (1969) believes that *Homo habilis*' ecological niches are similar to those occupied by *Australopithecus africanus* in South Africa.

Australopithecus africanus is excluded from the direct ancestry of *Homo habilis* as it is specialised in the direction of *Australopithecus robustus* and *boisei* (Johanson and White, 1979). They list four such specializations to support their contention:

- stronger molarization of the premolars,
- increased relative size of the post canine dentition,
- increased buttressing of its mandibular corpus, and
- increased robustness of the corpus itself.

White and co-workers (1981) add several more traits to this list; all being closely related to the hypertrophy of the posterior dentition. McHenry (1985) suggests that *Australopithecus africanus* immediately predates the appearance of *Homo habilis* and *Australopithecus robustus* plus *boisei*. It shares with these, the derived features that are absent in *Australopithecus afarensis*. Thus, *Homo habilis* evolved from a species with a complexity of traits associated with well-developed post canine megadontia (McHenry, 1985). *Homo habilis* is regarded as an undisputed representative of an evolutionary lineage parallel to that of the robust australopithecines (Rak, 1985). This is supported by the fact that the primitive topography of the face is preserved all along the *Homo* line to this day. The earliest *Homo*, in the form of *Homo habilis*, is so morphologically close to *Australopithecus africanus*, that it is difficult to separate them (Rak, 1985). The *erectus*-like morphology of the frontal bone is not seen in any of the *Homo habilis* crania.

TABLE 4.1
Morphological Features of the Cranium, Mandible, and Dentition of *Homo habilis* and *Homo rudolfensis*

	Homo habilis	*Homo rudolfensis*
Absolute brain size	X=610cc	X=751 cc
Relative brain size	IQ approx 4	IQ approx 4
Overall cranial vault morphology	Enlarged occipital sagittal contribution	Primitive
Endocranial morphology	Primitive sulcal pattern (Falk, 1983; Holloway, 1985)	Frontal lobe asymmetry (Falk,1983; Tobias,1987)
Suture pattern	Complex incipient	Simple torus absent
Frontal	supraorbital torus	Primitive
Parietal	Coronal> sagittal chord	Midface>upper face
Face — overall	Upper face> midface breadth	Breadth: markedly orthognathic
Nose	Margins sharp and everted: Evident nasal sill	Less marginal eversion: no nasal sill
Malar surface	Vertical or near vertical	Anteriorly inclined
Palate	Foreshortened	Large
Upper teeth	Probably two-rooted premolars	Premolars three-rooted; absolutely and relatively large anterior teeth
Mandibular fossa	Relatively deep	Shallow
Foramen magnum	Orientation variable	Anteriorly inclined
Mandibular corpus	Moderate relief on external surface Rounded base	Marked relief on external surface Everted base
Lower teeth	Bucco-lingually narrowed Post canine crowns M_3 reduction Reduced talonid on P_4 Mostly single rooted	Broad post canine crowns No M_3 reduction Relatively large P_4 roots, and 2T, or even twin, plate-like P_3 roots

Australopithecus continued to inhabit Africa during early Pleistocene, co-existing with more advanced hominids, such as *Homo habilis* ('Handy man'), until he became extinct about 1.75 million years ago. He walked erect and his brain, though smaller than ours, was bigger than that of an ape; he made simple pebble tools. It is not known today whether *Homo habilis* was no more than an advanced form of *Australopithecus* who has also become extinct, or whether he was an early ancestor of man.

Kamoya Kimeu discovered *Homo habilis* (KNM ER 1813) (Leakey, 1974). This specimen is similar to ER 1470, but is much smaller, with a brain size of 510 cc. Estimated age is 1.8-1.9 million years. Apart from its extremely small size, ER 1813 is quite similar to a number of *Homo erectus* and *Homo habilis* skulls. It is surprisingly modern, with a rounded skull, no sagittal crest, modest eyebrow ridges, and a small amount of nasal prominence.

Homo habilis is a controversial species, with much disagreement over which specimens belong to *habilis*, and which do not. A number of scientists now use the name *Homo rudolfensis* to refer to ER 1470 and some similar fossils. The smaller *habilis*-like specimens such as ER 1813 and ER 1805 are variously assigned to *habilis*, *Homo ergaster*, or to another as yet unnamed species. The name *Homo microcranous* has been proposed for ER 1813, but does not seem to be widely used. Wood and Collard (1999) have argued on theoretical grounds that *Homo habilis* and *Homo rudolfensis* should be moved into the genus *Australopithecus*.

TABLE 4.2
Cranial Characters of *Homo rudolfensis* and *Homo habilis*

Homo rudolfensis	*Homo habilis*
No supraorbital torus or supratoral sulcus (thickened superior orbital rim only)	Superiorly protruding supraorbital torus, bilaterally arched and thickest above mid-orbit, with supratoral sulcus
Frontal squama flat in coronal plane of post orbital constriction	Frontal squama convex in coronal plane of postorbital constriction
Superior facial breadth (Martin no. 43) less than mid-facial breadth (Martin No. 46)	Superior facial breadth greater than mid-facial breadth
less than mid-facial breadth (Martin No. 46)	mid facial breadth

(Contd.)

(Contd.)

Infraorbital plate inflated to coronal plane anterior to nasal bridge	Infraorbital plate that and vertical, in coronal plane at or posterior to nasal bridge
Anterior edge of maxillary zygomatic process over premolars and anterior to coronal plane of inferior orbital margin	Anterior edge of zygomatic process over P_4/M_1 or M_1, in coronal plane of inferior orbital margin
Preglenoid plane extensive and horizontal	Preglenoid plane restricted and steep
Molarized mandibular P_4.	Mandibular P_4 not molarized

See Wood and Uytterschaut (1987) and Wood (1992).

Some of the distinguishing features (Wood, 1992) are:

- ***Homo habilis*** is hominine with respect to its masticatory complex, has a relatively deep mandibular fossa, and retains an essentially australopithecine postcranial skeleton – bipedal hominid (1.9 million years ago). Features like absolute size of the maxilla, mandible, degree of parieto–sagittal curvature and molar crown size are intermediate between the australopithecines and the then oldest hominid taxon, *Homo erectus.*
- ***Homo rudolfensis*** is a mixture of a later *Homo*-like postcranial skeleton, face; and dentition is adaptively analogous to those of the robust australopithecines; has a large neurocranium with approximately 750 cc; is orthognathus and has a shallow mandibular fossa.

Homo sapiens are biologically close to the other higher primates, but it is only their behavioural characteristics that predominantly set the species apart. The most significant features of hominid behavioural evolution are the development of tool making capabilities, conceptual thought, and symbolic language. These features are relatively simple to examine in *Homo sapiens*, but in extinct hominids they can only be inferred indirectly from fossil remains.

Tool Making Capabilities

The ability to use tools is not unique to the Hominidae; vultures are reported to use rocks to crack open ostrich eggs, and sea otters use stones to open clam or oyster shells. It was long claimed that only hominids actually made tools, but in 1960 chimpanzees were observed in the wild breaking off twigs from trees, stripping away the leaves, and using the twigs to extract termites from their nests. Since the twig had to be modified by removing its leaves, this activity constituted tool making, even though at a crude level.

Tool use obviously preceded tool making in hominid development, but because the earliest tools were probably made from perishable materials, such as wood, it is not known when either of these activities became a regular feature of hominid behaviour. The first recognizable tools of hominids were made of stone, which, fortunately for archaeologists, were extremely durable. They have been found at sites with fossils, which date to at least two million years ago. By the Middle Pleistocene, the tools had developed into finely made, symmetrical, ovate hand axes that were flaked on both sides and carefully timed, and retouched to produce straight edges. These are the first standardised artifacts, (i.e., objects made systematically to conform to a pattern preconceived in the maker's mind) and are the first to show that the toolmakers' considerations were aesthetic as well as functional. Modern man has, of course, superseded stone tools with those made of metals.

Social Organization

The shift to terrestrial life in an open savannah environment would mean that a small, upright biped without the benefit of large, powerful jaws would be vulnerable to attack from predators. Opposing views have been proposed to explain why the early hominids not only survived but also prospered. One view is "the hunting hypothesis", which suggests that keen eyesight, cunningness, and the development of weapons necessary to hunt large animals channelled the natural aggression of humans and enabled them to defend themselves against predators. The other view stresses the significance of hunting as a stimulus toward social cooperation and food sharing.

Also important to the second view is the fact that human infants are much more helpless than other primate babies and require constant care for a relatively long period of time. With the formation of kinship groups a division of labour could occur, with the men doing the hunting and the women staying behind to gather plant foods and care for the children. Pair bonding within groups may have been established among the early hominids when the females became sexually receptive at all times, unlike other primate females who have an estrus cycle. Members of a group who perform different economic tasks during the day must operate within a limited territory and have a place where the entire group reassembles to distribute the food and spend the night. The concept of a home base has been confirmed in the archaeological record from early sites at Koobi Fora, where concentrations of stones and bones indicate places that were continuously or periodically occupied by groups for the purposes of butchering and eating animals and making tools.

5

Homo erectus

Fossil remains of what is now known as *Homo erectus* (Fig. 5.1) were first discovered in Java in 1891. Since this find, skeletal and cultural remains of *Homo erectus* have also been found in Africa as well as in Asia. The oldest *Homo erectus* fossils are from East Africa and date to about 1.9 mya. Most European fossils once referred to as *Homo erectus*, now seem to be more neatly fitted into an early *Homo sapiens* population. Beginning with *Homo erectus*, the story of human evolution is increasingly that of cultural elaborations. The *Homo erectus* tool assemblage is referred to as the Acheulian tradition (large hand axes).

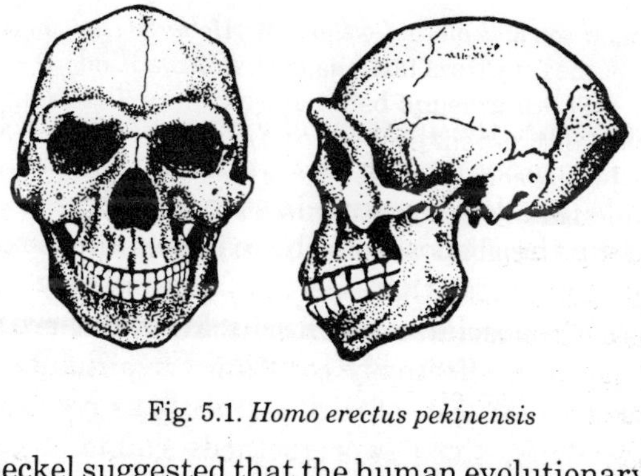

Fig. 5.1. *Homo erectus pekinensis*

Haeckel suggested that the human evolutionary line began among some extinct Miocene apes and reached *Homo sapiens* by way of an imagined speechless group of "ape-men" which he called *Pithecanthropi*. Dubois's "ape-man" walked erect as *Homo sapiens*. He christened his form as *Pithecanthropus erectus*, the 'erect ape-man'. (*Pithecanthropus* femur resembled that of modern man). Campbell (1962, 1965) opined that *Homo erectus* should include *Sinanthropus pekinensis* and *Pithecanthropus erectus*. The original generic name was subsequently dropped, and the fossils were renamed *Homo erectus* (subspecies designation *erectus* was assigned to distinguish it from other *Homo erectus* finds). *Homo erectus*, a synonym for archaic man or a primitive or an antiquated man, is a species of genus *Homo* immediately preceding us.

Skeletal remains of *Homo erectus* from Africa and Asia show moderate degree of individual and geographic variation. These forms comprise of a group whose morphological characters are sufficiently consistent and distinctive to justify their inclusion in the genus *Homo*. The *Homo erect*us skull is similar to the skull of modern humans (Fig. 5.2). The average brain size is larger than that of the habilines, and the skull balances on the spine in a way similar to ours.

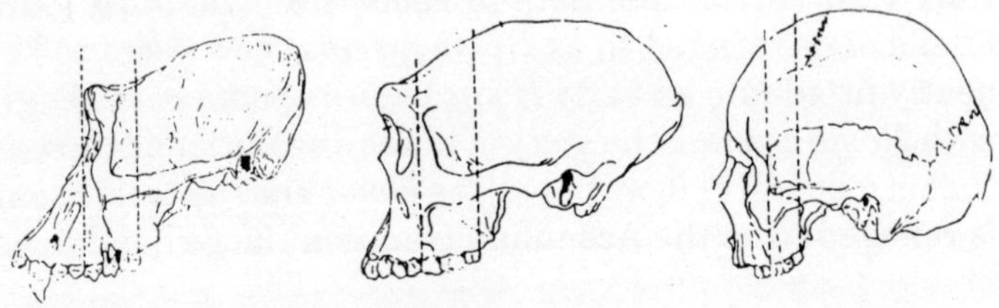

Fig. 5.2. Comparison of skulls of *Gorilla* (left), *Homo erectus* (middle) and Man (right)

To reconstruct the position of *Homo erectus* in hominid evolution, it is essential to define his place in time as precisely as possible. Modern developments in such disciplines as physics have placed at the disposal of the palaeoanthropologists, a variety of techniques that permit increasingly accurate assessments of the absolute age of fossils. Many of these methods are based upon the effectively constant (or absolute) rate at which radioactive isotopes of such elements as potassium and argon decay. When the newer methods cannot be applied, investigators may still ascribe a relative age to a fossil. This can be done by noting the contents of the layer of rock or the deposit in which the fossil was found; a layer containing, for instance, evidence of the remains primarily of extinct animals is probably older than one containing signs of predominantly recent or still living forms.

Such lines of evidence have led to the tentative conclusion that the species *Homo erectus* flourished over a long interval of Pleistocene time. The fossils recovered at the Koobi Fora sites may be about 1.6 million years old if radiometric dates obtained from associated volcanic materials are correct. Some of the remains from Olduvai Gorge are also ancient, and OH 9 is probably about 1.2 million years old. Many of the other older-appearing fossil hominids unfortunately lack absolute dates; but, by relative dating from several approaches, it would seem probable that the specimens from Sangiran and Modjokerto in Java (more especially the fossils from deposits of the Putjangan [Pucangan beds]) and perhaps from one of the Lan-t'ien localities in China are among the earlier representatives of *Homo erectus*. On the other hand, the youngest hominids accepted unequivocally as *Homo erectus* would seem to be the group from Ternifine in Algeria, Chou-k'ou-tien in China, and Trinil in Java.

Until more complete evidence is available, it is reasonable to suggest 1.6 million to perhaps 250,000 years ago as a time range for *Homo erectus*. For the most part, fossils dated earlier than this are the remains of *Homo habilis*. This species is best known from Olduvai Gorge and Koobi Fora in Africa, and the oldest specimens from there seem to be about 1.8 to two million years old. On the other hand, there is a group of later specimens that show some features of *Homo erectus* but are commonly regarded as transitional forms or as members of *Homo sapiens*; these include later Middle Pleistocene specimens from Europe (discovered at sites such as Bilzingsleben, Petralona, and Montmaurin), from North western Africa (Salé, Sidi 'Abd ar-Rahman, and Rabat), and from Asia (the Ta-li find of 1978). Other later forms suggest that *Homo erectus* had given rise to several regionally distinct forms, or subspecies, of archaic *Homo sapiens*, represented by late Middle Pleistocene or early Late Pleistocene fossils from Africa (Kabwe/Broken Hill, Elandsfontein [Hopefield/Saldanha], Cave of Hearths, Lake Ndutu, Omo, Bodo) and Europe (Swanscombe, Steinheim, Biache, Ehringsdörf, La Chaise).

Thus, the problem of recognizing populations as belonging to *Homo erectus* becomes more difficult; the boundaries of the species become blurred. These are the transitional zones in which a predecessor species seems to have been grading imperceptibly into its evolutionary product, *Homo erectus*, and in which *Homo erectus* apparently was undergoing further evolutionary change into its descendant species, *Homo sapiens*, to which modern humans belong.

TABLE 5.1
Cranial Characters of *Homo erectus* and *Homo habilis**

Homo erectus	*Homo habilis* "(*sensu stricto*)"
Thick cranial vault bones	Thin cranial vault bones
Tympanic plate thick and vertically deep: petrous crest strong	Tympanic plate thinner, shallow: petrous crest relatively weak
Nuchal torus[1] continuous between asteria thickened superiorly and inferiorly, with midline bulge	Prominent superior nuchal line or nuchal crest (not torus), weak laterally
Incipient to well-developed metopic (frontal) and sagittal (parietal) keeling: torus angularis[2]	Absent
Preglenoid plane extensive and horizontal	Preglenoid plane restricted and steep
Supraorbital torus[3] continuous with continuous, gutter-like supratoral sulcus.	Supraorbital torus bilaterally arched: supratoral sulcus divided bilaterally by locally thickened torus above midorbit
Index[4] on occipital scales> 100	Index on occipital scales<100
Nasal bridge strongly projecting	Nasal bridge projecting, but less prominent

*KNM-ER 1813, ER 3891, OH 13, OH 16, OH 24, OH 62, Stw.53.

[1]We follow Robinson's (1958. p. 398) distinction between nuchal torus and nuchal crest: A torus is a bony thickening, which is part of the skull architecture and is usually rounded, not sharply crested ..."[A torus is] not necessarily concerned with musculature except indirectly. Crests, on the other hand, are directly related to muscles..." A well developed nuchal torus is characteristic of Asian *Homo erectus*, and it is also present, albeit less conspicuously, on African specimens ER 730, ER 3733, Oh 9 and OH 12. Although the occipital of *Homo habilis* cranium ER 1813 possesses a rounded midline prominence, only a nuchal crest is developed laterally.

[2]A torus angularis is not present on ER 3733 or ER 3883, but it is found on OH 9. As described by Rightmire (1990), this feature also varies within the Asian sample.

[3]Supraorbital and supratoral form is essentially equivalent in ER 3883, Sangiran 17 and Sambungmachan. Other than the gracility of torus, the supraorbital region of ER 3733 is equivalent in morphology to that of Asian specimens and is different from that of *Homo habilis* (as described in this table). Both East African crania show incipient metopic keels. In superior view, the *Homo erectus* frontal squama is transversely convex behind the supratoral sulcus, and where the metopic keel is developed, it shows an anterior midline "peak". The convex frontal with a variably expressed "peak" is expressed on Asian as well as African crania here considered to represent *Homo erectus*. However, a convex frontal is also present in *Homo habilis* (ER 1813, Stw. 53, OH 16) and the majority of great apes: in *Homo rudolfensis* and *Australopithecus* spp. On the other hand, the squama is transversely flat or even slightly concave.

[4]The index is computed as inion-opisthion chord/lambda-inion chord x 100. Data on early hominids indicate that variation in this index is in part due to both gender and ontogenetic age (Kimbel *et. al.*, 1984). Thus, if ER 3883 is a male and ER 3733 a female, a relatively low index in the latter (Rightmire, 1990) should not be a surprise. Kimbel et al. also discussed the likelihood that a high index is produced in different taxa by dissimilar occipital growth patterns. Thus, an index value of greater than 100 in *Homo erectus* does not automatically qualify as a primitive character. This is contrary to the views of Andrews (1984).

Homo erectus is considered to be chronologically, morphologically, and genealogically intermediate between the lower Pleistocene African *Homo habilis* (Table 5.1; Fig. 4.1) and the later *Homo sapiens*. European specimens, however, indicate a *Homo erectus* to *Homo sapiens* transition. For this reason, there is considerable disagreement over how to classify the European fossils. *Homo erectus* was taller, larger brained, and more dependent on culture than the australopithecines but shorter and smaller brained than the *Homo sapiens*. Fossil materials illustrate graded series of morphological changes leading from *Homo erectus* to *Homo sapiens*.

Their life ways indicate use of fire, evidence of big game hunting, inhabited caves and built shelters. *Homo erectus*, in many physical and cultural characteristics, is intermediate between the australopithecines and the *sapiens* stage. Its cranial capacity ranges from 775 cc to 1225 cc with a mean cranial capacity of 1020 cc indicating that their skull is approximately twice the size of australopithecines, 50 percent larger than early *Homo*, and 75 per cent as large as *Homo sapiens* (Fig. 5.3). This increase in cranial capacity over time can probably be related to the increased body size over the time span of *Homo erectus* evolutionary history (Wolpoff, 1980).

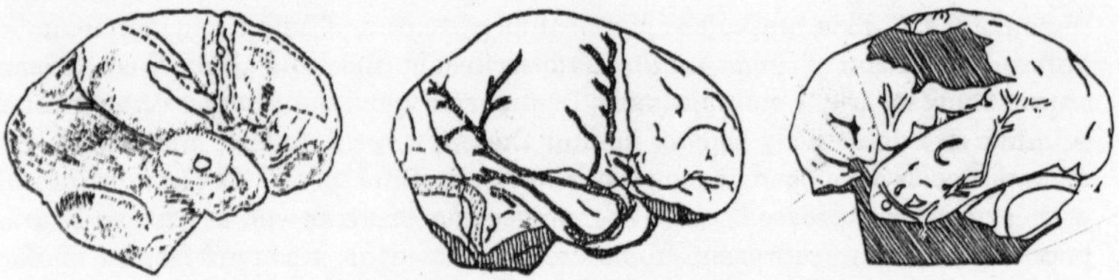

Fig. 5.3. Comparison of cranial casts of *Rhodesian* man (left), *Homo erectus* (middle) and Modern man (right)

Several characteristics on the skull of *Homo erectus* are immediately apparent: skull vault is flattened (platycephaly), reminiscent of a deflated football, i.e., long, low skull; overhanging the orbits is a pronounced supra-orbital torus, immediately behind it is the marked postorbital constriction, i.e., there is an indentation of the skull behind the large brow-ridges (Fig. 5.4). This area is elevated in modern humans because of an increased brain size. Occiput angles are directed sharply down and forwards, unlike the rounded profiles of the *Homo sapiens*. Protruding behind the occiput is the nuchal 'torus'. The forehead recedes instead of rising sharply as it does in modern humans.

Fig. 5.4. Comparison of the increasing prominence of the forehead associated with the decreasing postorbital constriction in a gorilla (left), *Homo erectus* (middle) and modern man (right)

The cranium of *Homo erectus* – with its low profile and average endocranial (interior) capacity of less than 1,000 cc (60 cubic inches) – is distinctly different from that of other humans. The average endocranial capacity of modern *Homo sapiens*, for example, is 1,450 cc, although the range for recent humans is 1,000 to 2,000 cc. The upper part of the range for *Homo erectus*, extending to more than 1,200 cc, overlaps with the lower values expected for *Homo sapiens*.

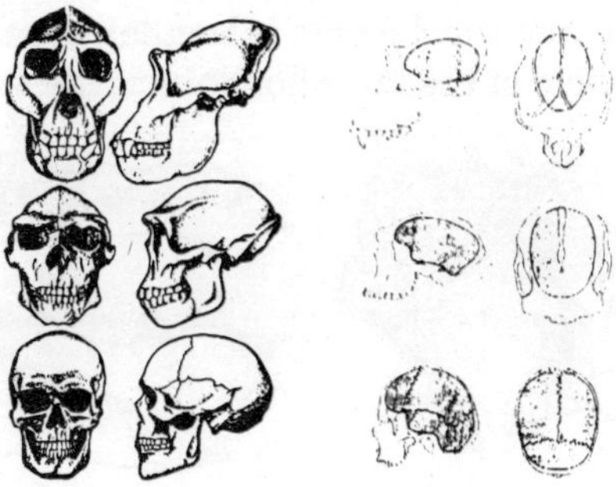

Fig. 5.5. Comparison of the skull profiles of *Gorilla* (above), *Homo erectus* (middle), and modern man (below). (Note the cranial capacity, forehead, supraorbital torus, shape of the skull, protrusion of the jaws)

Homo erectus skull is pentagonal in shape, is widest at or near its base unlike the skulls of modern humans, which are widest higher up on the vault – an important feature that distinguishes *Homo erectus* (Fig. 5.5). This could be due to thinner bones, reduced mass of the chewing muscles and the increased cranial capacities in modern man. A sagittal ridge highlights the skull's midline. *Homo erectus* skull bones are thick compared to those of *Homo sapiens*, and its nasal bones are flat. *Homo erectus* has alveolar prognathism, a heavily built mandible (Fig. 5.6), lacks a chin (receding chin) and the teeth are similar to but slightly larger than those of the modern human populations (Fig. 5.6, Fig. 5.7). These traits reflect heavy chewing. There are, indeed, a number of indications of well developed masseter muscles: the "buttresses on the facial skeleton – such as the broad nasal root, the thick supraorbital torus, the thick frontal processes of the zygomatic bone – and the well developed temporal lines" (Jacob, 1975). Heavy muscles require bony attachments and these can be seen in the ridges on the mandible. Differences in size and structure of teeth occur among the various *Homo erectus* forms (Fig. 5.6). In most of the earlier Javanese upper jaws, a diastema exists between the lateral incisor and the canine, and the canine projects a bit beyond the margin of the other teeth. These are non-hominid primate characteristics that are not found among most of the australopithecines. It is difficult to know what their adaptive function might have been. Characteristic of pongids and many australopithecines, too, is that the molar teeth increase in size from front to back. This sequence varies in *Homo erectus*, but in most cases, the third molar is smaller than the second; a design that is typical of the later hominid pattern.

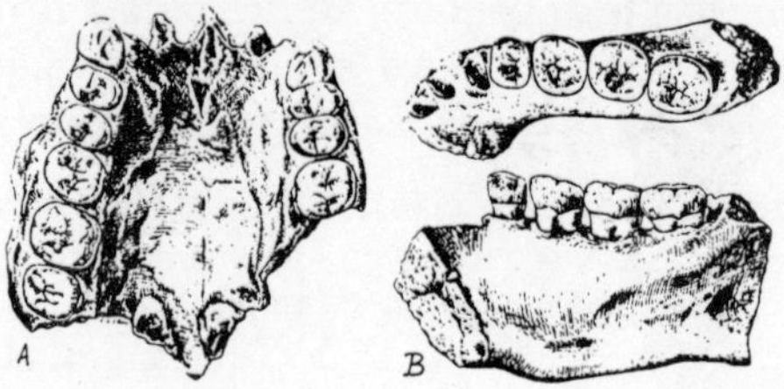

Fig. 5.6. *Homo erectus* [note the presence of diastema, size of the teeth, shape of the dental arches (A), and views of the upper and lower jaws (B)]

Two other dental traits of *Homo erectus* need to be mentioned:

- Shovel-shaped incisors – these have an adaptive response in hunting and gathering societies, such as the Eskimos, in which a great deal of chewing takes place with the anterior teeth. Additional enamel (on the raised ridges on both sides of the edges) may serve as structural support, which enables the teeth to last longer; this is specially pronounced in *Neanderthal* and some other modern populations.
- Taurodontism (or enlarged pulp cavity) – this provides an advantage among the hard-chewing humans and such grazing animals (e.g., horses, cows) that require high crowns on their teeth.

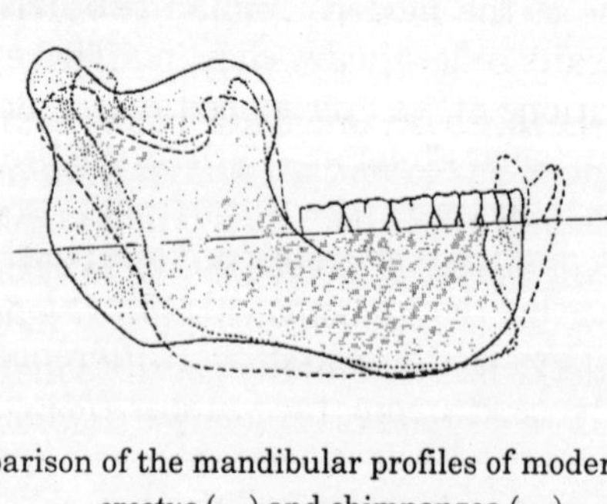

Fig. 5.7. Comparison of the mandibular profiles of modern man (…), *Homo erectus* (—) and chimpanzee (---)

Bodily Structure of *Homo erectus*

Much of the fossil material discovered in Java and China consists of cranial bones, mandibles, and teeth. The few broken limb bones found at Chou-k'ou-tien have provided little information. It is possible that the complete femur excavated by Dubois at Trinil is more recent in age than the other fossils found there and not attributable to *Homo erectus*. It comes as no surprise, therefore, that the greatest descriptive emphasis has been on the shape of the skull and not on the postcranial finds. The continuing discoveries in Africa – particularly at the Olduvai and Lake Turkana sites – have yielded a more complete picture of the anatomy of *Homo erectus*.

The femur is the most commonly recovered postcranial fossil. Apart from the Trinil specimen, the affinities of which are open to doubt, a number of femurs have been found at Chou-k'ou-tien, and more have been recovered from sites in Africa. Several notable characteristics have been observed in the *Homo erectus* femur. The form of these bones resembles that of modern humans, and *Homo erectus* must have walked upright efficiently. On the other hand, the construction of the bones is robust, a condition also seen in other skeletal members. This robusticity suggests that the life-style of *Homo erectus* was physically demanding.

Limb bones also supply information about the size of *Homo erectus*. Size is important to any hominoid species, since it influences behaviour and various aspects of anatomy, including bodily proportions. One measure of size is stature, or height. The femurs found at Chou-k'ou-tien and Koobi Fora are too broken to yield a good estimate of the height of these individuals, but accurate measurements of the skeleton numbered KNM-WT 15000 from Nariokotome have been made. Although he was not fully grown, it is thought that the Nariokotome boy would have reached about 160 centimeters in height, which is close to the stature expected for modern adult males.

Recent analysis suggests that there are a number of morphological differences in the locomotor skeletons of *Homo erectus* and *Homo sapiens*. The limb bones are essentially modern. Howell (1978) states that "Nonetheless, the functional locomotor capabilities appear to have been generally similar to adaptations in *Homo sapiens* for habitual erect posture and efficient striding bipedal gait". There is no doubt that *Homo erectus* was bipedal.

Its cultural associations, such as the Acheulian tool tradition, are elaborations of earlier cultural associations. Some morphological features distinguishing *Homo erectus* from earlier forms, such as increase in brain sizes, facial morphology, and perhaps dental structure, may be related to changes in culture and tool manufacture. Cultural elaboration and the physical evolution of *Homo erectus* are probably interrelated, although we are still unsure of exactly how the interrelationship functioned.

Broom and Robinson (1950) discovered at Swartkrans a fragment of mandibular corpus they claimed to be "of a type of early man which seems to us to be very near to that of *Homo*". This was found from the same breccia that was yielding remains of ape-man, *Paranthropus*, thus, demonstrating contemporaneity of the two forms of hominids. *Homo erectus* is a different kind of hominid which first appears in Africa about 1.8 mya and in Asia about 1.5 or 1.6 mya. It is the first hominid species to leave Africa and successfully inhabiting other regions utilizing greater technological skills, increasing cultural dependence, and such morphological features as a projecting nose. They were the first hominid species to show recognizably human adaptive patterns and show geographic distinctions that parallel those of *Homo sapiens*; they represent the evolutionary precursors of modern variations. *Homo erectus* is associated with one of the most momentous and far reaching human innovations, the controlled and purposeful use of fire. With this, humans began to shape the world according to their design. They used fire in caves, to cook food, perhaps in hunting, and to harden spears.

Homo erectus was an intelligent creature moulding its culture to meet new environmental demands and had a more complex cultural assemblage than its ancestor, *Homo habilis*. They were the first hominids to make tools according to set and regular patterns.

Researchers have uncovered fossils from Zhoukoudian Cave with large brains and other traits suggesting that this ancient population of China was evolving in a modern direction. Again, various details, notably the shape and orientation of the lower border of the cheekbones, link these fossils with the modern people of the region.

Origin of *Homo erectus*

Admittedly, the major problem confronting palaeoanthropologists today is an explanation for the origin of the *Homo erectus* fossils, which appear in the African records at or slightly before 1.5 mya. Some of the *Homo erectus* features (for instance, larger supraorbital and occipital torii, larger brain size, longer and lower cranial vault, nasal structure, and robust post-cranial morphology) are foreshadowed in the *Homo habilis* material already discussed. There are certainly important differences in robusticity and cranial thickness between typical *Homo erectus* fossils and African Pliocene hominids. It seems that some specimens referred to *Homo habilis* were already enlarged in body size with a greater post-cranial robusticity (Fig. 5.8). And perhaps, the *Homo erectus* cranial morphology is an allometric consequence of an enlarged body size.

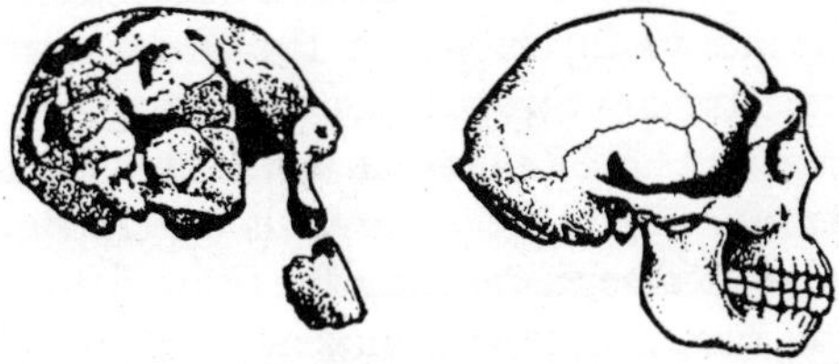

Fig. 5.8. Comparison of the skulls of *Homo erectus* (right) and *Homo habilis* (left)

An alternative explanation for the origin of *Homo erectus* is that the species developed outside the area with the best Plio-Pleistocene fossil record (East Africa). Evidence for the existence of hominids in Asia during the Pliocene is still poor, but claims for Yuanmon (China), Java and Ubediya (Israel) deserve further examination (Stringer, 1984).

Variations in *Homo erectus*

It has only recently become possible to assess the geographical variation in the early *Homo erectus* fossils. This certainly points to a greater morphological stability in the *Homo erectus* features in time and space than seemed likely when OH 9 was the only reasonably complete early African fossil representative of *Homo erectus*. Certain typical *Homo erectus* characteristics (e.g., occipital angulations, pelvic structure, etc) are found outside the *erectus* group and the supposedly "typical" characteristics (like the cranial thickness, sagittal keeling) are not universally present within it. As such, this makes it difficult to define the species morphologically by its own within-group traits.

Java man bears certain distinguishing characteristics, for instance, a robust cranium and a distinctively shaped face. Compared with *Homo erectus* material from elsewhere, this hominid species from Java has thick bones, pronounced supraorbital ridges and a well developed nuchal crest (shelf of bone at the back of the neck) for anchoring the neck muscles (Fig. 5.8). These early fossils from Indonesia have large projecting faces with massive rounded cheekbones and large teeth. Their face bears a number of small but important features: a "rolled edge" on the lower margin of the orbits, a distinctive ridge on the cheek bones, and a nasal floor that merges smoothly with the face. These and other characteristics combine to create a special Indonesian variation within the *Homo erectus* group.

Some scientists have also proposed splitting *Homo erectus*. The Turkana Boy and ER 3733 fossils would then become *Homo ergaster* (Tattersall, 1993). *Homo erectus* would have a large average brain size than *ergaster*, and the brow ridges may have a different shape, flaring out to the side more (Burenhult, 1993).

Some difference in estimated brain size is apparent between the Javanese and the Chou-k'ou-tien populations of *Homo erectus*. Thus, for seven Javanese craniums, the average is 883 cc, with a range from 750 to 1,030 cc; while for five Chou-k'ou-tien craniums, the capacity ranges from 915 to 1,225 cc and averages about 1,043 cc. That is, the mean capacity in the Peking fossils of *Homo erectus* exceeds that of the Javanese by about 160 cc. Investigators note with interest that the cranium from Kung-wang-ling (also Chinese), which comes from an earlier period than do those of Chou-k'ou-tien and is an approximate contemporary of the earlier fossils from Java, shares with the Javanese group a smaller cranial capacity (780 cc). Theoretically, the difference in brain size between the two groups of Asian fossils of *Homo erectus* may be the consequence of further evolutionary increase in brain size in later populations of *Homo erectus*. Alternatively, it may simply be interpreted as representing differences in average features between two different "races" or subspecies of *Homo erectus*. Several African values are available, and in the case of the Koobi Fora and Olduvai individuals, these values vary from about 850 cc (ER 3733) to 1,067 cc (OH 9).

While the cranial capacity of *Homo erectus* falls short of that of *Homo sapiens* by about 570 cc on the average for the Javanese group and about 400 cc for the Peking group, *Homo erectus*, in turn, exceeds the australopithecine capacities by an average of about 440 cc in the case of the Javanese fossils and about 600 cc for the Peking group. Thus, the 'cerebral gap' between *Australopithecus* and *Homo erectus* is slightly greater than that between *Homo erectus* and *Homo sapiens*. Into the former group fit the cranial capacities of the group of fossils known as *Homo habilis*. Clearly, the last word has not been written on their affinities. The possible evolutionary place of the cranial capacity of *Homo erectus* can be seen in Fig. 5.5.

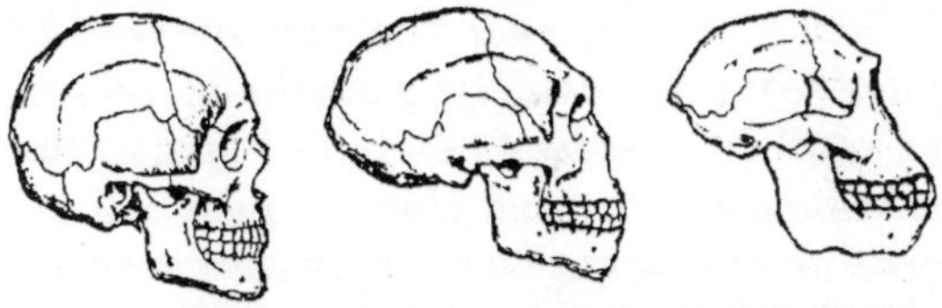

Fig. 5.9. Comparison of skulls of *Australopithecus* (right), *Homo erectus* (middle) and Modern Man (left)

Apart from their characteristically small capacity, the skulls of *Homo erectus* show a number of distinctive features. The face, which is preserved in only a few specimens, is massively constructed and projecting in its lower parts. The bone forming the wall of the nose is thinner and more everted than in earlier *Homo* or *Australopithecus*, and the nasal profile is not too different from that seen in modern humans. The braincase is low, with sides that taper upward, and the bones of the cranial vault are thick. Over the eye sockets is a strongly jutting brow ridge, called a supraorbital torus. There is a flattened forehead, and the front part of the cranium immediately behind the supraorbital torus is appreciably constricted from side to side. A low ridge or keel of bone may extend from the frontal bone onto the parietals (side walls) along the midline, and there tend to be strongly developed crests in the mastoid region. The broad-based skull has a sharply curved occiput and a markedly thickened shelf of bone (occipital torus) that divides the upper and lower occipital surfaces. The area of neck-muscle attachment below the occipital torus is much larger than in *Homo habilis* or *Homo sapiens* (Fig. 5.10). Other distinguishing features in *Homo erectus* can be found on the underside of the cranium, especially at the joint of the mandible. The mandible itself is deep and robust and lacks chin development. The teeth are on the whole larger than those of *Homo sapiens*.

Transition from *Homo erectus* to *Homo sapiens*

The supposed transition from *Australopithecus* to *Homo erectus* via *Homo habilis* provides very little support for a gradual evolution between Pliocene and Pleistocene hominids. On the other hand, the transition from *Homo erectus* to *Homo sapiens* seems to be represented by genuine mosaics occurring at the right time and place. This establishes that the evolutionary intermediates do occur over a major part of the known Pleistocene range for the two species. The plausible explanation for these widely distributed mosaics seems to represent another lineage of hominids living in the same areas as that of *Homo erectus* and *Homo sapiens*, displaying mixed characters, or they might be hybrids between the two species; the extent of temporal or geographic overlap between the two species is, however, difficult to establish. Perhaps the *Homo erectus* – "early *sapiens*" – transition was gradual and widespread, characterized by mosaicism, whereas the appearance of *Neanderthals* and archaic modern *Homo sapiens* was punctuational and localized, leading to easy recognition of taxonomic boundaries.

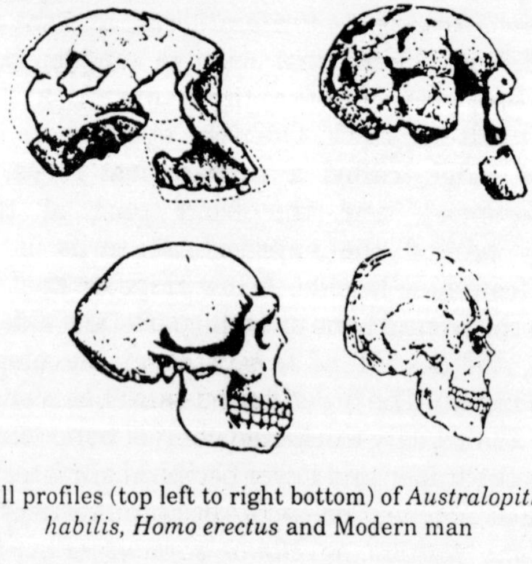

Fig. 5.10. Skull profiles (top left to right bottom) of *Australopithecus*, *Homo habilis*, *Homo erectus* and Modern man

The total pattern of the body structure of *Homo erectus*, as preserved in the bones, is rather different from that of *Homo sapiens*. Parts of the postcranial skeleton are robust but otherwise generally comparable to those of modern humans. The brain is relatively small, though not so small as that of *Australopithecus* and *Homo habilis*. In addition, in this hominid's thick skull bones and extraordinarily developed eyebrow ridges and occipital torus, some investigators say they see unique, specialized features, not characteristic either of its presumed ancestors or of apes and not pointing to *Homo sapiens* as the direction of subsequent evolution (Fig. 5.10).

Some scientists even infer that these last traits show *Homo erectus* to have specialized so far off the modern human line that it could not have been ancestral to *Homo sapiens*. It is notable that the more ancient *Australopithecus* had thin skull bones and only modest protuberances on the cranium; while the later *Homo sapiens* also tend to have thin skull bones with a marked diminution in the size of crests and ridges sculpted on the surface of the cranium. It is often said that only two choices are open in interpreting this situation: either *Homo erectus*, with a thick cranium and large adornments on the skull, could not have been on any direct evolutionary line from *Australopithecus* to *Homo sapiens*; or else *Homo erectus* evolved specialized features from *Australopithecus* and then lost them again (or underwent a type of evolutionary reversal) to produce the thin, smooth cranium of modern *Homo sapiens*.

Yet, the two choices offered for solving the problem are judged by some authorities to be oversimplified. It has been observed that the exponents of either view probably fail to take into account that there is very little evidence about the variability of these features – cranial thickness and external embellishments of the skull – among members of even one population of *Homo erectus*, let alone among different populations of *Homo erectus* dispersed through two or three large continents. Then, too, practically nothing is known about the climatic or ecological conditions under which cranial thickening occurred, or of the effect on skull growth of the brain enlargement that was so striking a feature of the evolutionary advance of earliest Pleistocene humans to the people of the later Middle Pleistocene and of the Late Pleistocene. These and many other questions need answers before

Homo erectus can be written off as an ancestor of *Homo sapiens*. In the meantime, another hypothesis that meets most of the available evidence is that *Homo erectus* was in the process of evolving from pre-*Homo erectus* – probably *Australopithecus* and *Homo habilis* – to post-*Homo erectus*; that is, to *Homo sapiens*. In most details, the bodily structure of *Homo erectus* fulfils what might have been predicted for an intermediate between *Australopithecus* and *Homo sapiens*.

Other *Homo erectus* Finds

Based on the few post-cranial bones, including the fragments of partial skeletons, discovered in 1986 at the Olduvai Gorge, it is well known that *Homo habilis* was built much like an australopithecine in overall body size and limb proportions. *Homo erectus*, in contrast, was the first really big hominid. Tall, thin, and barrel-chested, about 14 per cent bigger and 20 per cent heavier than *habilis* and was well adapted to long-distance ranging on the African savannas.

Narmada Man

This is the first *Homo erectus* material found in India from the right bank of the river Narmada Son in Hathnora in Madhya Pradesh (Central India) (Sonakia, 1982) and comprises of a nearly complete and well-preserved cranium, with its face missing. This was an *in situ* find on the surface of an alluvial terrace of late middle Pleistocene (about one million years ago). This elongate cranium is believed to be that of a female about 30 years of age. Its maximum cranial breadth is situated low, at the level of the temporals; has an occipital angulation; an increased cranial capacity estimated to be between 1155 and 1421 cc; less marked post-orbital constriction; greater frontal expression than in the classic *Homo erectus*; and relatively high expansion of the temporal squama. All these characteristics suggest that the *Narmada* Man belongs to a group of more evolved *Homo erectus*. Its norma occipitalis is pentagonal in form. On the other hand, the marked basion-bregma height (beyond that recorded for either *Homo erectus* or *Homo neanderthalensis*) with relatively less receding frontal bone than in the other *Homo erectus* remains is difficult to interpret. Like middle Pleistocene Europeans, the 'Indian *Homo erectus*' possesses characteristics that can be assigned to both *Homo erectus* and *Homo sapiens*.

- *Homo erectus* traits: presence of supraorbital torus, sagittal keel, thick vault bones, greatest width at the level of the temporals – low on the skull.
- *Homo sapiens* traits: presence of an increased cranial capacity, less marked post-orbital constriction, and lack of a sharply angulated occipital.

Particularly, because of its '*sapiens*' features, *Narmada* Man has been referred to as an "evolved" *Homo erectus* (de Lumley and Sonakia, 1985). Associated with the cranium were mammalian bones and tools of Acheulian industry, especially the hand axes and cleavers.

Homo erectus, pretty well known above the neck, is the most primitive human or the prehuman fossil from the terminal Pliocene or early Pleistocene epoch. This ape-like cannibalistic being of low intelligence, but with a braincase more or less like that of modern man, represents the earliest man-savage interface. *Homo erectus,* which is broadly comparable in his level of cultural development to some of the more primitive races of mankind still surviving, displays both primitive and advanced features. They existed about 500,000 years ago and probably were ancestral to the modern humans.

Fossil remains of a teenage boy, 1.5 million year old *Homo erectus* specimen, have been reported from the site of Nariokotome, West of Lake Turkana in Kenya. This is the most complete skeleton known of all fossil hominids.

Heidelberg Man (*Homo heidelbergensis*)

This was discovered by gravel pit workers near Heidelberg in Germany in the year 1907 ; its estimated age is between 400,000 and 700,000 years. This find consisted of a lower jaw with a receding chin and all its teeth. The jaw is extremely large and robust, like that of *Homo erectus*, but the teeth are at the small end of the *erectus* range. It is, therefore, identified as *erectus* on the basis of its age, but could be an archaic *sapiens* as well.

Early archaic *Homo sapiens* (*Homo heidelbergensis*), with cranial capacities around 1150 to 1250 cc (similar to that of modern humans), appeared in Europe and other parts of the Old World during 300,000 and 200,000 years ago. They were low-browed individuals having robust skeletons. Their cranial vault and thickness of the bones suggest an intermediate status between *Homo erectus* and modern humans. Evidence for transition from *Homo erectus* to early archaic *Homo sapiens* has been discovered from numerous sites in Africa, Hungary, Germany, Greece, Spain, France, Italy, Indonesia and China. *Neanderthals* succeeded the European forms of the early archaic men about 150,000 years ago.

Asian Fossils

Subsequent discoveries gradually established the case for a new and separate species of fossil hominids. At first, these discoveries were centred largely in Asia. At several different places in Java, essentially similar fossils were found: the sites other than Trinil are Kedung Brubus, Modjokerto, Sangiran, Sambungmatjan (Sambungmachan), and Ngandong. Another series of finds was made in China, especially in the caves of Chou-k'ou-tien (Zhoukoudian) near Peking, although virtually all of the remains from there subsequently were lost during the Sino-Japanese War late in 1941. Newer discoveries have since been made in the caves, while three new Chinese sites – at Kung-wang-ling (Gongwangling) and Ch'en-chia-wo (Chenjiawo) in the Lan-t'ien (Lantian) district of Shensi province, and at Hohsien (Hexian) in Anhwei province – have yielded remains attributable to *Homo erectus*.

By the end of World War II, the pattern of early discovery had given rise to an idea that *Homo erectus* was a peculiarly Asian expression of early humans. Subsequent discoveries in Africa served to change this view, and it came to be realized that Europe, too, might have harboured *Homo erectus*.

African Fossils

In Africa in 1954-55, excavations at Ternifine, East of Mascara, Algeria, yielded remains who's nearest affinities seemed to be with the Chinese form of *Homo erectus*. Other hominid fragments from north-western Africa – parts of a skull found in 1933 near Rabat in Morocco and jaws and teeth from Sidi 'Abd ar-Rahman (Sidi Abderrahman, 1954) in Morocco – show features reminiscent of *Homo erectus*, though they are rather more advanced in structure than those of Ternifine and of Asia. Another fossil probably related to *Homo erectus* is a cranium found in 1971 at Salé, Morocco. Although nearly all of the face and part of the frontal bone have been broken away, it is an important specimen.

Some of the more convincing evidence for the existence of *Homo erectus* in Africa came with the discovery in 1960 of a partial braincase at Olduvai Gorge in Tanzania. This fossil, called OH 9, was picked up by L.S.B. Leakey in deposits of Bed II, the second of four numbered beds (or layers) identified in the Olduvai stratigraphic sequence, beginning with Bed I (the oldest) and ending with Bed IV (consisting of more recent sediments). Additional cranial remains, jaws, and limb bones of *Homo erectus* were later discovered in Beds III and IV and the overlying Masek levels.

Much of the Olduvai material is fragmentary, but gaps in the knowledge of *Homo erectus* in East Africa have been filled to some extent through finds made by Richard Leakey. Since 1970, a number of valuable fossils have been unearthed at localities on the eastern shore of Lake Turkana (Lake Rudolf) in North-western Kenya, now commonly referred to as the Koobi Fora sites. Included in these assemblages are the remains of *Australopithecus* and probably some representatives of the species *Homo habilis*. Of several specimens that clearly do belong to *Homo erectus*, one cranium from Koobi Fora (KNM-ER 3733) is quite complete and well preserved; it is likely to be one of the most ancient *Homo erectus* fossils discovered anywhere in Africa. Other significant finds at Koobi Fora include a nearly intact skeleton (ER 1808), although it has been shown to have come from a diseased individual. Another nearly complete skeleton (designated KNM-WT 15000) found at Nariokotome (West Turkana), a site on the north-western shore of Lake Turkana, is of an adolescent male.

European Fossils

The realization that Africa as well as Asia were apparently peopled by a form of mankind classifiable as *Homo erectus* led to a re-examination of some of the earliest hominid fossils from Europe. An isolated mandible (lower jaw bone) had been found in a sandpit just north of Mauer, close to Heidelberg, Germany, in 1907. Although it had been given a variety of names over the years, its exact affinities to other fossils remained uncertain, because no associated cranium was found. Since then, a number of investigators have come to regard the Mauer mandible as representing a member of the species *Homo erectus*. Although its geologic age is perhaps comparable with that of the Chou-k'ou-tien hominids in China, this skeletal fragment from Europe shows more modern structural features than do the Asian and African jaws of *Homo erectus*. The exact significance of these features in the Mauer jaw is still being debated; they could be the mark of an individual variant, highlighting the fact that it is not yet known how variable in bony structure *Homo erectus* was. Alternatively, the Mauer specimen could represent a subspecies or race of *Homo erectus* that is slightly more advanced in anatomic structure than are the African and Asian populations; or it is possible that the Mauer individual could have been a member of a very early population of *Homo sapiens*. Fossils discovered in Hungary have provided evidence for the latter view.

In 1965, remains of two individuals – a child and an adult – were found in a travertine quarry in Hungary, about 30 miles (50 kilometers) west of Budapest. The remains of the child are milk teeth, and enough of them are preserved to show affinities with the Chinese *Homo erectus* of Chou-k'ou-tien. The adult is represented by a large part of an occipital bone from a large-brained skull. While showing some features reminiscent of *Homo erectus*, the general form and the capacity of the Hungarian cranium also suggests an affinity with an early branch of *Homo sapiens*. In fact, it appears to be related to somewhat later Middle Pleistocene skulls (perhaps 200,000 years old) from Europe, such as those of *Swanscombe* and *Steinheim*, which are accepted as early members of *Homo sapiens*. Since the *Vértesszollos* group of remains exhibits such twofold affinities, authorities differ as to whether to call the population they represent *Homo erectus* or *Homo sapiens*. The same uncertainty applies to some of the fossils found in North Africa; the Sidi 'Abd ar-Rahman and Rabat remains are regarded by some experts as late surviving members of *Homo erectus* and by others as forms transitional between *Homo ere*ctus and *Homo sapiens*.

The European fossils that are of equivalent age to those of Chou-k'ou-tien and the later north-western African localities – including the jaw from Mauer; the teeth and occipital bone from *Vértesszollos I*; and some skull parts, teeth, and post-cranial remains discovered more recently at Arago cave (near the village of Tautavel in south-western France) – appear to have more structural features in common with early *Homo sapiens* than do their Asian and African contemporaries. This is also true of several other European finds that are somewhat younger (that is, of later Middle Pleistocene antiquity). Pieces of braincase and an upper molar tooth from travertine deposits at Bilzingsleben, Germany, show some resemblances to *Homo erectus*, but the skull fragments are more comparable to those of a remarkably complete hominid recovered in 1960 from a limestone cave near Petralona, Greece. Detailed study of the anatomy of this latter hominid suggests that it shares with *Homo sapiens* some traits that are not characteristic of *Homo erectus*. There is still no firm consensus regarding the classification of these later Middle Pleistocene populations of Europe. It may be safest to conclude that the case has not yet been convincingly established for the existence in Europe of *Homo erectus* as known from Asia and Africa.

Behavioural Inferences

To understand this species of fossil humans more completely, it requires looking past their bones and beyond their dispersal in time and space. What evidence is there that bears upon the behavioural or cultural pattern of their daily existence?

First, the discovery sites themselves may throw light on this question. At Chou-k'ou-tien, the remains of *Homo erectus* were found in cave deposits; this in itself does not prove that these hominids were consistent cave dwellers. But the additional evidence of associated remains of stone and bone that seem to have been accumulated by these creatures – charred animal bones, collections of seeds, and what could be ancient hearths and charcoal – all point to *Homo erectus* as having spent appreciable periods of time as a troglodyte (cave dweller) at Chou-k'ou-tien. The remains of Lan-t'ien, Trinil, Sangiran, and Modjokerto, as well as Ternifine and Olduvai, were all found in open sites, sometimes in stream gravels and clays, sometimes in river sandstones, in conglomerate and volcanic rocks, or in lake beds. These suggest that *Homo erectus* also lived in open encampments along the banks of streams or on the shores of lakes; proximity to water was crucial to human survival. The remains of the *Vértesszollos* individual showed that he occupied mud flats deposited along with minerals around springs in a tributary of the Danube River. These open presumed campsites revealed by excavation contain abundant stone implements and stone chips that seem to have resulted from manufacture, fractured and partly burnt bones of animals that could have been hunted for food, and traces of what appears to have been a hearth.

Thus, both Chou-k'ou-tien and Vértesszollos have shown signs that Middle Pleistocene humans had a controlled mastery of fire. Indeed, outside a cave at Chou k'ou-tien, charcoal was found along with traces of a stone tool-making industry in an open gully deposit that seems to be slightly older than the cave deposit itself (containing the bones of *Homo erectus*). In this region of China, therefore, it has been observed that the earliest convincing indication of the use of fire by humans immediately precedes the earliest example of a cave being occupied by them. This supports the notion that successful cave dwelling by human creatures depended on their first having mastered fire.

It was not only cave dwelling that mastery of fire seems to have made possible, and able to keep warm, man was able to move into colder climes; indeed, this factor might have speeded the migrations of ancient humans into the chilly, often glaciated regions of prehistoric Europe. Sooner or later, too, humans started cooking their food, thus, reducing the work demanded of their teeth. This, in turn, may have played an important part in minimizing the evolutionary advantage of big teeth – cooked food needs far less cutting, tearing, and grinding than does raw food. This relaxation of the evolutionary selective pressure that favoured the survival of people with strong, big teeth may, in turn, have led directly to a diminution in the size of the teeth – one of the features that has come to distinguish *Homo sapiens* from *Homo erectus*.

Other signs of the culture of *Homo erectus* are the implements found in the same deposits as their bones. Chopping tools made from split pebbles characterize both the Chou-k'ou-tien and Vértesszollos deposits; both are members of a so-called chopper-chopping-tool family of industries. At Ternifine, in north-western Africa, *Homo erectus* was found in association with totally different kinds of stone implements; these comprise bifacial hand axes and scrapers that have been characterized as representing what archaeologists call an early Acheulean industry. This is part of the great Acheulean hand-axe complex of human industry, remnants of which are found widely spread over large parts of Europe and Africa. An Acheulean industry is known also from Olduvai Gorge, as is a local, more ancient form of stone chopper manufacture known as the Oldowan industry; but the exact cultural associations of these stone tools with African *Homo erectus* (as exemplified by OH 9) are uncertain.

Hence, *Homo erectus* has been found associated in some parts of the world with a chopper-chopping-tool tradition and in other places with an Acheulean bifaced hand-axe industrial complex. Numerous non-hominid animal bones also occur along with the remains of *Homo erectus*; sometimes these bones seem to have been broken deliberately or burned. From this evidence, it seems that *Homo erectus* was a hunter. His bodily (including cerebral) endowments and manufactured equipments were so much superior to those of *Australopithecus* and *Homo habilis* that it is highly probable that his food-collecting techniques, including hunting, also were better. Many scientists hold that *Australopithecus* and *Homo habilis* were more scavengers than hunters, perhaps at best opportunistic hunters who seized their chance when a weak, young, sick, or aged animal crossed their paths. Indeed, many of the animal bones found in australopithecine deposits are of juvenile and aged creatures; and, although larger animal bones have been recovered from *Homo habilis* deposits, these have exhibited tooth marks of non-hominid predators as well as cut marks. *Homo erectus*, on the other hand, seems to have been a confirmed hunter, and his prey included animals of all age groups.

It can credibly be supposed that, as with present-day hunters such as the Kalahari San (Bushmen) and the Australian Aborigines, meat from the hunt formed only a part of the diet of *Homo erectus*. Other juicy morsels may have been furnished by snakes, birds and their eggs, locusts, scorpions, centipedes, tortoises, mice and other rodents, hedgehogs, fish, crustaceans, and numerous other edible forms of life. Even children could have caught many of these — as they do in the Kalahari today, before they are allowed to accompany the older men on the hunt. Vegetable food also must have played a big part in the diet of *Homo erectus*, in the form of fleshy leaves, fruits, nuts, and roots. Accumulations of hackberry seeds, for example, were found in the Chou-k'ou-tien cave deposits.

There seems little doubt that *Homo erectus* must have been

omnivorous (as *Homo sapiens* is today), for such a diet is the most opportunistic of all, and modern humans are the most opportunistic of all living primates. Emancipated from too narrow an environmental dependence, from too restricted a dietary regimen, humans have come to live off many diets, in many surroundings. *Homo erectus* was probably one of the earliest of the great opportunists; and it is likely that his very opportunism endowed the species with evolutionary flexibility, with adaptability, and with a very elastic survival kit.

Another question that may be asked about *Homo erectus* culture is whether there is any evidence of ritual among these extinct people. There is no sign that they buried their dead; no complete burials have been found, no graves, no grave goods, and no red ochre (a mineral used as a paint by later forms of hominids) on or around the bones. That cannibalism was practiced was once inferred from the Chou-k'ou-tien finds, but little evidence remains to support such a hypothesis.

Evolutionary Implications

A few workers have generally opposed the view that *Homo erectus* was the direct ancestor of *Homo sapiens*. L.S.B. Leakey argued energetically that *Homo erectus* populations, particularly in Africa, overlap in time with more advanced *Homo sapiens* and, therefore, cannot be ancestral to the latter. His arguments were based primarily on stratigraphic considerations and chronology, but some support for Leakey's point of view has come from the analysis of anatomical characteristics exhibited by the fossils. By emphasizing the distinction between "primitive" and "derived" traits in the reconstruction of relationships between species, several palaeontologists have attempted to show that *Homo erectus* does not make a suitable morphological ancestor for *Homo sapiens*. Because the braincase is long and low, is thick-walled, and presents a strong supraorbital torus, they claim *Homo erectus* to show derived (or specialized) characters not shared with more modern humans. At the same time, it is noted that *Homo sapiens* do share some features – such as a rounded, lightly built cranium – with earlier hominids (e.g., the smaller, or "gracile," *Australopithecus* species or *Homo habilis*). It is clear that these palaeontologists consider the early, gracile Plio-Pleistocene *Homo habilis* to be more closely related to *Homo sapiens* than is *Homo erectus* (a conclusion with which Leakey would have agreed).

These findings have not been widely accepted. Stratigraphic arguments concerning the overlap of (late) *Homo erectus* and (early) *Homo sapiens* are no longer compelling, and it is difficult to reject hypotheses of continuity between Middle and Late Pleistocene *Homo* populations on morphological grounds. When the effects of body size on scaling of the brain are taken into account, for example, it is apparent that gracile australopithecines (and *Homo habilis*) have relatively rounded skulls because they are small-bodied creatures. General resemblances between these earlier hominids – with their more spherical, lightly constructed craniums – and *Homo sapiens,* therefore, cannot be used to demonstrate a direct relationship. Instead, studies of size and scaling in human evolution have emphasized that representatives of *Homo* can be grouped into a reasonable ancestor-descendant sequence showing increases in body size through time. Despite having a heavier, more flattened braincase, *Homo erectus* is not out of place in this sequence, and there is no reason to expect that hominid craniums should have remained sub-spherical in form, as bodies grew larger in the Middle Pleistocene.

If this much is agreed, there is still uncertainty as to how and where populations of *Homo erectus* evolved into early *Homo sapiens*. This is a major question in the study of human evolution, and one that resists resolution even when hominid fossils from throughout the Old World are surveyed in detail. Several general hypotheses have been advanced, but there is still no firm consensus regarding models of gradual change (local evolution in different geographic regions) as opposed to scenarios of rapid evolution in one region followed by migration of the new populations into other areas.

Gradualistic Views of the Transition to *Homo sapiens*

A traditional view held by many anthropologists is that a species may be transformed gradually into a succeeding species in the same lineage. Such successive species in the evolutionary sequence are called chronospecies. The boundaries between chronospecies seem almost impossible to determine by any objective anatomic or functional criteria; what is left is the guesswork of "ruling off" at a moment in time to draw the boundary. Thus, competent authorities have seriously suggested that, in the last analysis, such a chronological boundary may have to be drawn arbitrarily between the last survivors of *Homo erectus* and the earliest members of *Homo sapiens*. The problem of defining the limits of chronospecies is not peculiar to *Homo erectus*; it is one of the most vexing questions in palaeontology.

Such gradual change with continuity between successive forms has been postulated particularly for North Africa (where *Homo erectus* at Ternifine is seen as ancestral to later populations at Rabat, Temara, Jebel Irhoud, and elsewhere) and for Southeast Asia (where *Homo erectus* at Sangiran may have progressed toward populations such as those at Ngandong and at Kow Swamp in Australia). Some workers have suggested that similar developments could have occurred in other parts of the world. Certainly, the most thoroughly documented statement of this view has been provided by the American anthropologist Carleton S. Coon (1962) in *The Origin of Races*. Coon departed from traditional gradualist theory when he postulated the transition from *Homo erectus* to *Homo sapiens* as a crossing of a kind of evolutionary Rubicon – that is, a clear-cut boundary, or threshold, marked not only by bodily changes but also by striking cultural changes.

This supposed interrelation of cultural achievement and the shape and size of teeth, jaws, and brain is a theorized state of affairs with which many students of fossil hominids strongly disagree. Throughout the human fossil record, there are examples of dissociation between skull shape and size, on the one hand, and cultural achievement, on the other. A smaller-brained *Homo erectus* from China seems to have been among the first humans to tame fire, while much bigger-brained people in other regions of the world, living later in time, have yielded no evidence that they knew how to handle fire.

The core of Coon's thesis is that *Homo erectus* evolved into *Homo sapiens* not once but five times, as each subspecies of *Homo erectus*, living in its own territory, passed the postulated critical threshold. Since this part of Coon's theory depends on accepting his supposed *erectus-sapiens* threshold as correct, it is opposed by those who find the threshold concept at variance with the modern genetic theory of evolutionary change.

A corollary of this concept was Coon's suggestion that different subspecies (or "races") of man evolved at different rates and, hence, crossed the *erectus-sapiens* threshold at different times. In particular, Coon has suggested that "the step from the ancestral *Homo erectus* to the modern *Homo sapiens* was taken by Caucasoid men in Europe no less than 200,000 years before the same step was taken by Negro men in Africa." Not only was this part of Coon's reasoning based on a concept that is at odds with modern evolutionary theory, but also some of the assumed "facts" on which it was based are highly problematic. He claimed that the races of today had already appeared before the *erectus-sapiens* transition. Apparently by allocating two very early fossil human skulls from Europe (Swanscombe from England and Steinheim from Germany) to the European, or Caucasoid, "race," although they lived 250,000 years ago, he concluded that the supposed *erectus-sapiens* transition had occurred among Europeans, or Caucasoids, as long as 250,000 years ago. Few, if any palaeoanthropologists would agree with Coon that these early skulls are to be recognized as those of members of the Caucasoid, or European, "racial" group, as generally understood.

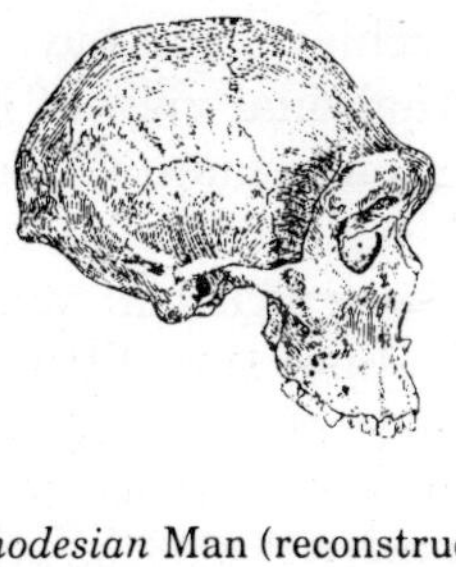

Fig. 5.11. *Rhodesian* Man (reconstruction of skull)

Turning to Africa, Coon claimed that "*Rhodesian* man" [represented by parts of several skeletons found in Zambia (formerly Northern Rhodesia) in 1921] belonged to *Homo erectus*. Coon, who thought that *Rhodesian* man (Fig. 5.11) lived only 30,000 or 40,000 years ago, thus argued that *Homo erectus* was still present in Africa long after he had disappeared from the European scene. This argument hinges on controversial judgments of the species to which *Rhodesian* man is to be assigned and on dating of the fossil remains. In contrast to Coon, most palaeoanthropologists regard *Rhodesian* man as a member of an African subspecies of *Homo sapiens*, commonly dubbed *Homo sapiens rhodesiensis*. Also, it is now thought that the bones are of later Middle Pleistocene age. The evidence for the late survival of *Homo erectus* in Africa is questionable at best.

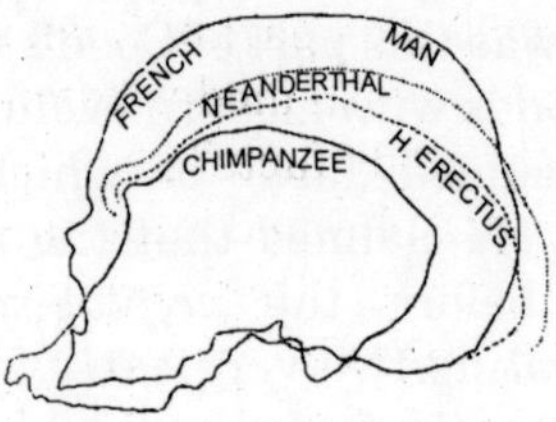

Fig. 5.12. Superimposition of Craniums of Modern man (outer), *Neanderthal*, *Homo erectus* and Chimpanzee

Alternative Modes of Species Change

That phyletic gradualism marks the transition from *Homo erectus* to *Homo sapiens* (Fig. 5.12) is one interpretation of the fossil record, but the available evidence can also be read differently. Some workers who do not hold Coon's ideas of body change tied to cultural achievement have come to accept what can be termed a punctuational view of human evolution. In their view, species are seen as more stable entities, not simply as arbitrarily defined segments of a lineage. It has been suggested that species such as *Homo erectus* may be expected to exhibit evolutionary stasis (i.e., little or no morphological change) over long periods of time. The transition from one species to a descendant form may occur relatively rapidly in geologic terms and in a restricted geographic area rather than on a worldwide basis. Whether this may be a correct interpretation of the situation involving *Homo erectus* and *Homo sapiens* is unclear; that is, whether *Homo sapiens* evolved from populations of more archaic *Homo erectus* gradually, or rapidly during a short "pulse" of evolution late in the Middle Pleistocene, has not been settled, although a good deal of attention has been given to the problem.

The continuation of the argument underlines the need for more fossils to establish the range of physical variation of *Homo erectus* and for more discoveries in good archaeological contexts that permit increasingly precise dating. Additions to these two bodies of data may settle remaining questions and bring the unsolved problems of phylogeny and classification of *Homo erectus* nearer to resolution.

6

Neanderthal Man

In 1856, workers quarrying for limestone in the Neander Valley near Düsseldorf, Germany came across a skull and some limb bones. In the succeeding years, many other specimens were found, not only in the Neander Valley, but also in countries such as France, England, Italy, and Iraq and as far south as Israel. Controversy surrounded the interpretation of these fossils. German Anatomist Rudolf Virchow examined the first discovery and concluded that it was a *Homo sapien* with rickets, caused by a Vitamin D deficiency. He also theorized that his flattened head was due to powerful blows. As more finds were made, also with the appearance of rickets, this was considered too coincidental and they were now considered sub-human.

Fig. 6.1. *Neanderthal* men

In the early 1900s, after many skeletons were found, the French palaeontologist Marcellin Boule, determined that *Neanderthals* could not fully extend their legs, walked stooped over, and had their head thrust forward (Fig. 6.1). This notion would be the popular image for about fifty years.

In 1957, researchers re-examined the skeleton Boule had

examined and concluded that *Neanderthals* walked upright and that the stooped posture suggested by Boule's specimen was due to a case of arthritis. More evidence from various sites has shown that *Neanderthals* "wielded simple tools, wore body ornaments, had religious rites and ceremoniously buried their dead". Today he is classified as totally human – *Homo sapien.*

Neanderthal man, type of early human, existed 125,000-35,000 years ago and was generally considered a subspecies of *Homo sapiens*. The *Neanderthals' middle* Palaeolithic culture included stone tools, fire, burial, and cave shelters. The so-called classic *Neanderthals* were robust and had a large, thick skull, a sloping forehead, a chinless jaw, and a brain somewhat larger than that of modern humans; they stood slightly over 5 ft (152 cm). It is unclear whether *Neanderthals* were replaced by *Homo sapiens sapiens* or interbred with other early humans.

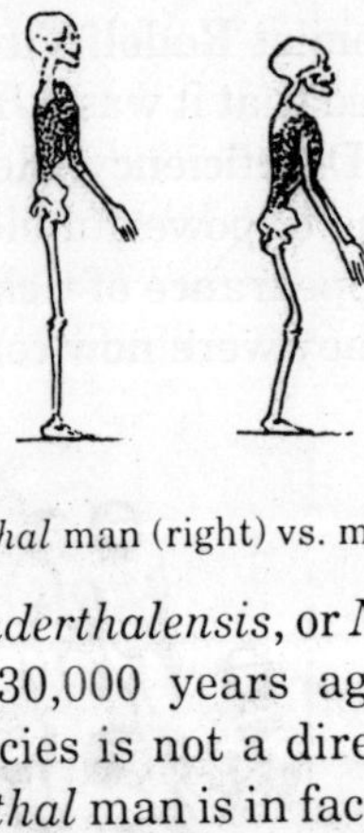

Fig. 6.2. *Neanderthal* man (right) vs. modern man (left)

Homo sapiens neanderthalensis, or *Neanderthal* man, lived between 230,000 and 30,000 years ago. While it has been concluded that this species is not a direct ancestor of modern man (Fig. 6.3), *Neanderthal* man is in fact quite similar to them. Some characteristics of *Neanderthal* man mimic *Homo erectus*, such as the protruding jaw and sloping brow. *Neanderthal* man's brain size is a bit bigger than modern man, at an average of 1450 cc (Fig. 6.3). *Neanderthals* mostly lived in cold climates, and their body proportions were similar to those of modern cold-adapted people: short and solid, with short limbs. The average height was about 5'6" (168 cm), and by bone mass it is believed that they would seem extraordinarily strong by today's standards.

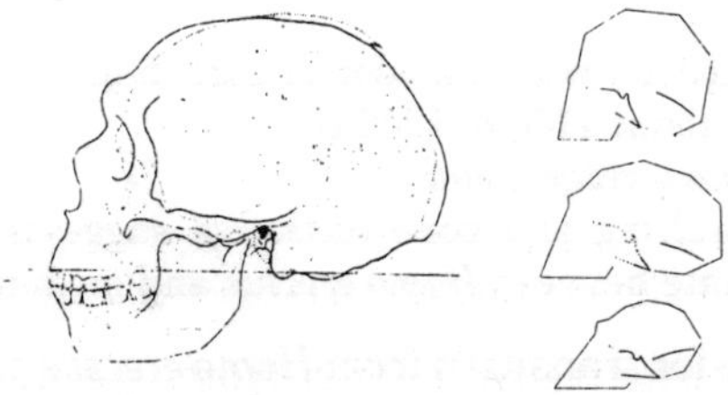

Fig. 6.3. Craniums of *La Chapelle* and modern man (superimposed left; note sloping protrusion of jaw, sloping forehead and brain size) and their geometric representations (right; top-man; middle-*La Chapelle*; bottom-Chimpanzee)

Neanderthal man was well advanced, using a wide variety of tools with precision. This species is also noted as the first species that buried their dead, and also relied on a more evolved social structure than past *Homo* species.

The Extinction of *Neanderthal* Man

About 40,000 years ago, *Cro-Magnon* man ("modern" man) lived in Africa and the Middle East, while Europe was the home of *Neanderthals*. Both were *Homo sapiens* (meaning they could probably interbreed). *Neanderthals* were short, stoop-shouldered, and had only basic tools. *Cro-Magnon* men were taller, more slender, and used more tools.

Scientists used to believe that 40,000 years ago the *Cro-Magnons* swept through Europe and quickly wiped out the *Neanderthals*. Recent findings indicate that the extinction of *Neanderthal* man was not so fast as this.

A cave on the Mediterranean coast has revealed that *Neanderthal* men were living there only 30,000 years ago. This is confirmed both by radio carbon dating and thorium-uranium dating. This gives good evidence that *Cro-Magnon* and *Neanderthal* lived side-by-side for about 10,000 years.

So why did *Neanderthals* vanish? Were there race wars? Did the populations interbreed (diluting the *Neanderthal* genes)? Were the *Cro-Magnons* better at competing for scarce resources? We can only speculate.

Archaic and Modern Peoples

Early Archaic *Homo sapiens (Homo heidelbergensis):*

- 300 to 35 kya; between 300,000 and 200,000 years ago hominids with nearly modern cranial capacities appeared in Europe and other portions of the Old World. viz. robust skeletons;

- cranial capacity within modern range (but at the low end of the range) — about 1150 to 1250 cc;
- had low-brow ridges; and
- cranial vaulting and bone thickness suggests that they were intermediate between *Homo erectus* and modern humans.

Evidence for transition from *Homo erectus* to Early Archaic people has been discovered from numerous sites:

- **Africa:** Bodo, Awash Valley, Ethiopia; Kabwe (Broken Hill), Zambia; Ndutu, Lake Ndutu, Tanzania; Sale, El Hamra, Morocco.
- **Europe:** Vertesszollos, Hungary; Swanscombe, England; Bilzingsleben and Steinheim, Germany; Petralona, Greece; Arago, Spain; Biache, France; La Chaise, Italy.
- **Southeast Asia:** Ngandong (Solo), Indonesia; and Dali, China.

Neanderthals succeeded European forms about 150,000 years ago.

However, it is not yet known as to what were their linguistic and symbolic abilities (Table 6.1) and their relationship to modern humans. They reveal many morphological changes, viz. more gracile cranial anatomy, smaller face, smaller teeth, smaller jaws, more rounded skull, and considerably less robust post-cranial skeleton (Table 6.2).

TABLE 6.1
Symbolic Behaviour in the Middle Palaeolithic

Very little consistent patterning is seen for *Neanderthal* burials. The burials are rare and idiosyncratic (each unique). One possible pattern is the tendency for males to have tools and females not to have tools. There is problem with recognizing pits/burials (soil stains) and interpreting the evidence.
Evidence of non-utilitarian artefacts is extremely limited at the Mousterian sites.
Tata, Hungary (Mousterian site): ivory plaque (charm).
Shanidar Cave, Iraq: cranial deformation is seen in two crania. This perhaps suggests desire for beauty?
Haua Fteah, Cyrenaica, Libya: bone whistle (or flute); dates to about 60,000 years ago.

During the last 100,000 years, there appear to have been two different kinds of humans, viz. *Neanderthals* and modern *Homo sapiens* (Table 6.2). *Neanderthals* appeared in Europe during 15,000-100,000 years ago; the youngest

Neanderthal skeletons date to 36,000 to 31,000 years ago in Western Europe. Anatomically, modern *Homo sapiens sapiens* appear in Africa about 100,000 years ago. The *Neanderthals* appear to be cold-adapted humans found in Europe, Western Asia, the Near East and North Central Asia (Table 6.3).

TABLE 6.2
Comparison of Important Characteristics of *Neanderthal* and Modern *Homo sapien sapiens*

Features	*Neanderthals*	*Modern H.s. sapiens*
Posture	Same	Same
Manual abilities	Same	Same
Range and characteristics of movement	Same	Same
Cranial capacity	Same (Mean 1450 cc; is slightly higher than that in modern man	Same
Skeletal framework	Short and powerful; thick, slightly curved long bones; large areas for muscle attachment	Considerably less robust post-cranial skeleton
Skull	Bulging occipital bun; long, low skull without a prominent forehead; large cheek bones; prominent brow ridges; mid-facial prognathism, i.e., projected nasal area; receding or virtually absent chin	More gracile cranial anatomy; smaller face; more rounded skull
Teeth and jaws	More robust; larger front teeth showing extreme wear (used as vices); strong masticatory apparatus; larger jaw (large enough to accommodate 3rd molars-wisdom teeth)	Smaller; relatively small front teeth

The *Neanderthals* (Cave Men), probably the best known prehistoric humans, comprise of a highly variable population inhabiting diverse habitats. Morphologically speaking, they include populations, which are indistinguishable from the modern man in postcranial skeleton (Table 6.2; 6.3). The long bones are more robust in the cave man, as a whole, suggesting that he was a well-muscled individual. The *Neanderthal* man is often referred to as a squat, shaggy, burly, beetle-browed hairy fellow with a sloping forehead (Fig. 6.4) and a slouching back. He had a fully upright posture (Fig. 6.2) with a cranial capacity (1,300 to 1,640 cc) equal to or greater than modern man (850 to 1,700 cc) but a facial skeleton heavier than that of modern man (Fig. 6.4).

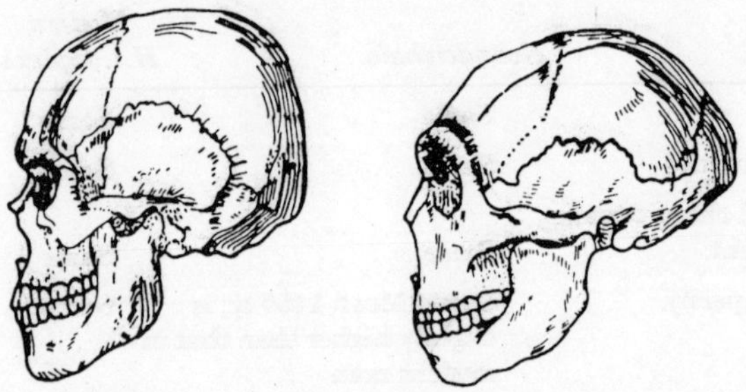

Fig. 6.4. Comparison of the craniums of modern man (left) and *Neanderthal* man (right)

Till the early 1970s, *Neanderthals* were lumped with other archaics as having modern brains and large primitive faces (and teeth). Western European *Neanderthals*, whose faces were longer and more projecting, and whose elongated heads appeared to have an "occipital bun" of bone at the back (Fig. 6.5), were simply more extreme than the others.

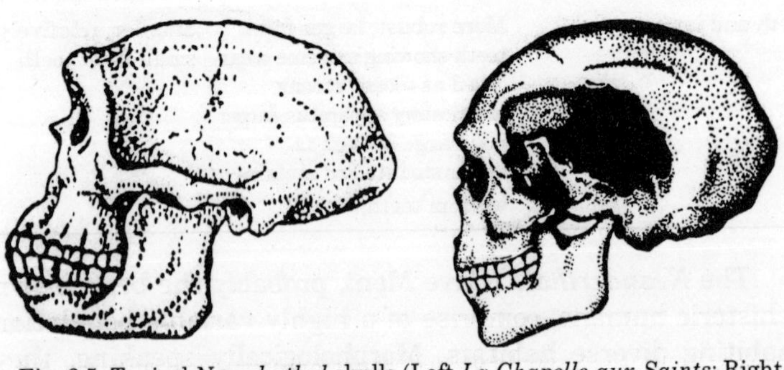

Fig. 6.5. Typical *Neanderthal* skulls (Left-*La Chapelle-aux-Saints*; Right-*Combe Capelle*)

TABLE 6.3
Evolution of Early Humans

	Homo habilis (small)	*Homo habilis (large)*	*Homo erectus*	*Archaic Homo sapiens*	*Neanderthals*	*Early Modern Homo sapiens*
Height (m)	1	1.5	1.3 - 1.5	?	1.5 - 1.7	1.6 -1.85
Physique	Relatively long arms	Robust but 'human' skeleton	Robust but 'human' skeleton	Robust but 'human' skeleton	Adapted for cold climate	Modern skeleton; adapted for warm climate
Brain size (cc)	500 - 650	600 - 800	750 - 1250	1100 - 1400	1200 - 1750	1200 - 1700
Skull form	Relatively small face; nose develo-ped	Larger, flatter face	Flat, thick skull with large occipital and brow ridges	Higher skull; face less protruding	Reduced brow ridges; thinner skull; large nose; mid-face projection	Small or no brow ridges; shorter, high skull
Jaws/ teeth	Thinner jaw; smaller, narrow molars	Robust jaw; large, narrow molars	Robust jaw in larger individuals; smaller teeth than H. habilis	Similar to H. erectus but teeth may be smaller	Similar to archaic H. sapiens; teeth smaller except for incisors; chin develop-ment in some	Shorter jaws than Neanderthals; chin developed; teeth may be smaller
Distribution	Eastern Africa	Eastern Africa	Africa, Asia, Indonesia	Africa, Asia, Europe	Europe and western Asia	Africa and western Asia
Known date (ya)	2-1.6 million	2-1.6 million	1.8-0.3 million	400,000-10,000	150,000-30,000	130,000 - 60,000

The principal effect of the new dating techniques has been the establishment of the great antiquity in Europe of the *Neanderthal*-type long face – big nose, and the flattened bulge at the back of the head. The oldest fossil now referred to as *Neanderthal* (Le Biache, France) was discovered in 1976 and is about 190,000 years old, while other fossils (f.i., Arago in the Pyrenees) with some *Neanderthal* characteristics, date to the 300,000s or older. Newer, more precise radiocarbon dates from the end of *Neanderthal* times show that, in some specific sites, the transition from *Neanderthal* to *Cro-Magnon* was quite abrupt. *Neanderthal* from St. Caesar in France, found in 1979, is about 35,000 years old, whereas the *Cro-Magnon* fossils probably date to at least 34,000, based on comparisons with the Pataud site in the neighbourhood. Such an abrupt transition does not leave enough time for evolution to have occurred in place. Besides, the oldest modern human fossils and archaeological sites of the Aurignacian culture of *Cro-Magnon* (Fig. 6.6) are found in eastern Europe just before 40,000 years ago, while *Neanderthals* still lived in the west. Modern humans from Qafzeh, in Israel, in the Near East itself, excavated in the 1960s, have been dated to ca. 92,000 years ago by thermoluminescence on burnt flints.

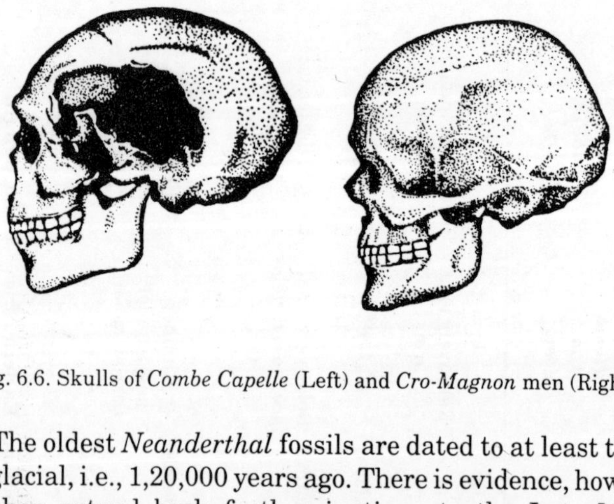

Fig. 6.6. Skulls of *Combe Capelle* (Left) and *Cro-Magnon* men (Right)

The oldest *Neanderthal* fossils are dated to at least the III Interglacial, i.e., 1,20,000 years ago. There is evidence, however, that they extend back further in time, to the Late Middle Pleistocene; the earliest fossils of *Neanderthal* man being no more than 2,50,000 years old at the most. Though, the *Homo sapiens sapiens* first appeared about 50,000 ya, yet this 'stone age man' (*Neanderthal* Man) can be considered a subspecies of the early *Homo sapiens*, i.e., *Homo sapiens neanderthalensis*. The *Neanderthal* man, which was still around when *Combe Capelle* appeared (Fig. 6.5), disappeared from the fossil record between 50,000 and 40,000 ya, i.e., they were gone by the IInd stadial of the IV Glacial.

It is suggested that *Neanderthals* were the cold adapted archaics, and the earliest modern humans were tropical. This shifting pattern implies that the distribution of the two populations was originally limited by ecological considerations; the Near East represented a boundary zone that shifted as the world's climate changed.

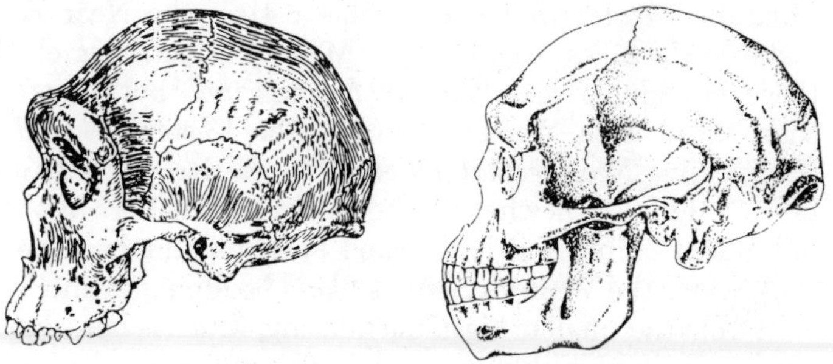

Fig. 6.7. Skulls of *Rhodesian* (Broken Hill) (left) and Solo (right) men

Neanderthals, like other cold-adapted animals, had very large deep chests and short lower arms and legs (Fig. 6.1), to better conserve body heat. New studies of the face suggested that the very long projecting face and huge, broad nose were distinctive; other large-faced archaics from Africa or East Asia had shorter, flatter faces, with more angulated cheekbones. Discovery of a pelvis from Kebara, in Israel, suggests that the way the body was carried was quite different, as the spinal column was more deeply indented into the back than in us.

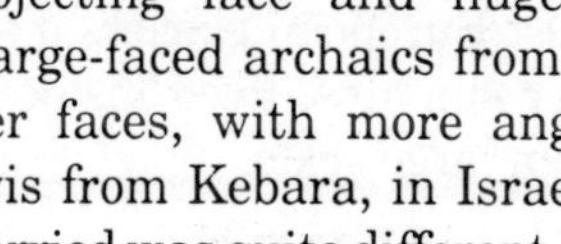

Fig. 6.8. Mandibular profiles of *Neanderthals* (left-*Tabun*; right-*Combe Capelle*)

Numerous fossils of the *Neanderthal* man have been reported from in and near Europe, middle and the Far East, Asia, Africa and Russia (Table 6.4; Figs. 6.7; 6.8). However, the key finds are from Dordogne in France – the site from where many of the early *Homo sapiens sapiens* finds have been reported. No fossil evidences have been recovered from America and Australia as man did not inhabit both until modern man appeared on the scene there.

Neanderthals of Western Europe became extinct during 40,000 to 50,000 ya and were replaced by people like us. The general feeling is that the Western *Neanderthals* failed to adjust and/or cope with the glaciers as periods of milder climate were too short for the type of the evolutionary process that led to the evolution of the modern man. Most of the evidence for this change is evident from the *Neanderthal* sites (Table 6.4) in the Near East; for instance, the caves on the slopes of Mount Carmel in Israel. Though these individuals were definitely *Neanderthals* but were different than those from Western Europe; the Eastern *Neanderthals* (Progressive *Neanderthals*) were characterized by less massive brow ridges, somewhat rounded skull, smaller faces, more prominent chins, and longer and straighter limbs. One of the reasons for the Western *Neanderthals* becoming extinct was the shift from the small to big game hunting, which is more intensive, continuous and highly organised exploitation of the big game. Besides, big game hunting called for co-operation on a band basis. It seems that the *Neanderthals* of Western Europe had not yet sufficiently evolved to be able to live together in large groups through long periods of severe cold climates; deep snows further restricted their movements and they were, thus, forced to disperse into smaller or single family units. But *Neanderthals*, who lived far away from the West European glaciers, i.e., in the Southeast, had relatively warmer climates, lived inside the caves and had group co-operation on a large scale.

TABLE 6.4
***Neanderthal* Sites**

Country	*Region / Site*	*Dating (years ago)*
Riss-Würm (III Interglacial)	Period	
France	Fontechévade	
Germany	Ehringsdörf	

(Contd.)

Yugoslavia	Krapina	
III Glacial Period		
	Kanjera	
	Florisbad	41,000 - 37,000
CENTRAL AND EASTERN EUROPE		
Yugoslavia	Vindija	41,000 - 32,000
Czechoslovakia	Ochoz	
	Sipka	32,000 - 28,000
	Ganovce	
	Kulna	45,000 - 38,6000
	Sala	
Hungary	Subalyuk	
WESTERN EUROPE		
France	La Ferrassie	75,000 - 60,000
	La Moustier	58,000 - 42,000
	La-Chapelle-aux-Saints	75,000 - 60,000
	Hortus	
	Saint Cesaire	34,000 - 31,000
	La Quina	
East Germany	Taubach	Middle Palaeolithic
West Germany	Wildsheuer	Early Würm
	Salzgitter-Lebenstedt	55,600
Bulgaria	Baco Kiro	43,000
Gibraltar	Forbes Quarry	Early Würm
		Levallois-Mousterian
Belgium	Spy	Early Würm
	Montmaurin	
	Dordogne	
Italy	Monte Circeo	
MIDDLE EAST		
Israel	Tabun	80,000 - 50,000
	Skhul	45,000
	Kebara	62,000 - 52,000
	Qafzeh	92,000
Iraq	Shanidar	60,000 - 46,000
ASIA		
China	Changyang	
	Da-li	71,000 - 41,000
	Ma-pa	2,50,000 - 1,28,000
II INTERGLACIAL		
Europe/Germany	Steinheim	
England	Swanscombe	
	Saccopastore	
	Teshik-Tash	

Generally, this variable group of *Neanderthals* could be divided into two main categories (Table 6.5): Progressive and Classic *Neanderthals* (Fig. 6.9; 6.10). The Progressive *Neanderthals* mostly consist of the earliest European fossils from the III Interglacial (warm interglacial) and are more like the modern *Homo sapiens* in their skeletal characteristics: lighter

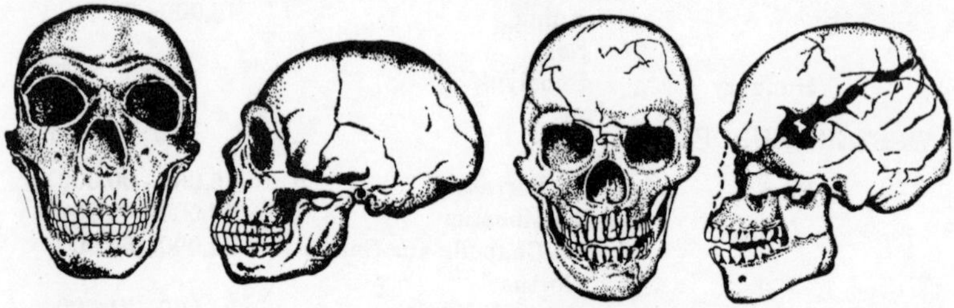

Fig. 6.9. Skulls of Classic (left) and Progressive (right) *Neanderthal* men

bones, hollowed cheeks, slight brow ridges, absence of occipital protuberance, etc. These features are quite similar to the later *Combe Capelle* Man. The later fossils from the IV Glacial in Europe, on the other hand, possess more extreme and rugged Neanderthaloid characteristics (Classic *Neanderthals*). This stadial population corresponds to the *Cro-Magnon* Man in their possessing more robust skeletal structure.

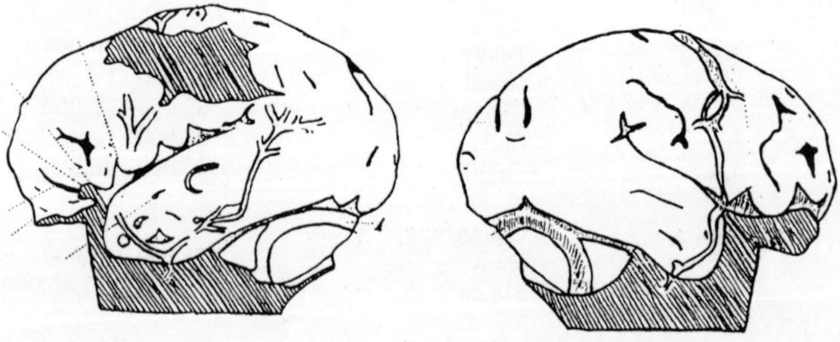

Fig. 6.10. Endocranial casts of *La Chapelle aux Saints* man (left-left side; right-right side).

Regarding this two-fold division of the *Neanderthals* (Table 6.5), one opinion is that the Progressive *Neanderthals* were ancestral to the modern man while the Classic *Neanderthals* were over specialized and represented the evolutionary dead end. The second opinion is that these differences between the two groups were not as clear cut as they were first thought to be; some of the differences could be due to faulty reconstructions of the poorly preserved fossil material. Therefore, the differences within these two groups could probably be due to population variation.

TABLE 6.5
The Progressive and Classic *Neanderthals*

Progressive *Neanderthals* *Es-Skhul* V (Mount Carmel)	high cranial vault; large and rugged lower face; large and rugged brow ridges (but clearly reduced from the typical *Neanderthals*) which are more or less indistinguishable from modern man; prognathus face; full and rounded occipital region; represents the evolutionary transition from the *Neanderthal* to the modern man.
Classic *Neanderthals* *La-Chapelle-aux-Saints*	adult male skull, large and robust in built; well preserved; low vault; large face; receding forehead; large, heavy supra-orbital arches; voluminous orbits; broad nose; stout mandible; no chin; small mastoid processes.
Tabun I (Mount Carmel)	fragmented skull; small, low vaulted cranium; skull breadth is maximum towards the back; pronounced development of supraorbital torus; no bun formation but instead a marked occipital torus is present; small mastoid processes; orthognathus face.
Circeo	adult male skull; teeth missing; sloping forehead; heavy brow ridges; large face; protruding occipital region.

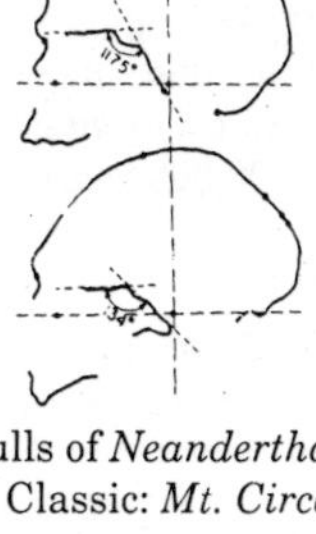

Fig. 6.11. Comparison of skulls of *Neanderthals* (top – Progressive, bottom – Classic: *Mt. Circeo)*

Burials

The *Neanderthal* burials are rare, each one of them is unique and shows very little consistent pattern (Table 6.6.). One possible pattern evident is the tendency for male burials to have tools whereas the female burials do not have tools. At times, it is even difficult to recognize pits.

TABLE 6.6
Burial Sites of *Neanderthal* Man

France

body placed in shallow trench; bison leg placed on chest; trench filled with bones, tools, and other debris.

La Ferrassie

cemetery (?) containing man, woman, two children, and two infants;

pits with human remains:

adults - all buried with east-west orientation

male buried in extended position with large flat slab of rock placed on the chest; and woman in flexed position; two or three mounds cover infant burials.

pits without human remains:

decay?; post-abandonment carnivore disturbance?

Italy

Guattari Cave, *Monte Cicero*

burial found in isolated inner chamber of cave:

skull with hole cut in base placed inside a circle of stones; skull was upside down and the right side of the skull was crushed by violent blows; near the skull were three piles of bones: red deer, cattle and pigs.

Siberia (Himalayan foothills)

Teshik-Tash

child buried in grave surrounded by six goat skulls whose horns had been jabbed into the ground.

Iraq

Shanidar Cave

chamber 110 m by 70 m; ceiling 14 above modern surface; deposits 15 m deep;

stratigraphy:

Layer A — 0 to 7,000 years ago,

Layer B — 7,000 to 12,000 years ago,

Layer C — Upper Palaeolithic deposits with abundant roof fall, and

Layer D — Middle Palaeolithic deposits; 9 m thick

burials dated as 35,000 to 100,000 years ago; contain deep ash deposits.

Middle Palaeolithic burials:

seven adults and two infants:

(i) 40 yr-old male killed by roof fall (not buried); withered humerus (arm amputated above elbow or perhaps useless from birth), arthritis, blind in one eye;

(ii) adult male killed by roof fall; covered by more stones by people; fire build over body;
(iii) adult male killed by roof fall; ribs scarred by wooden spear prior to death;
(iv) four individuals in a single grave; two adult females, one adult male, one infant;
(v) two individuals show cranial deformation; and
(vi) an adult male buried with garlands of flowers. (What evidence is there for the flower offering with this burial is not quite clear).

Israel

Kebara Cave

Qafzeh, Israel

Physical Characteristics

The *Neanderthals* have a long, large, low and wide skull with a flat crown and the maximum width is further back on the skull than in the modern men (Table 6.7; Fig. 6.12). They often have a projecting occipital protuberance, especially in the European fossils, heavy brow ridges forming a continuous bony arch over both the eyes (supraorbital torus) but not quite as pronounced as in *Homo erectus* (Fig. 6.12). The forehead rises but not as steeply as in the modern man. The face projects forwards in the middle and has a high or prognathus nose. Their jaws are large and heavy and lack the pointed chin as in modern man. The teeth in the *Neanderthal* man tend to be larger than in modern man. However, both of them have the same vertical face and arched tooth row accompanied by the identical forms of the teeth. Post cranially, the *Neanderthals* are short and powerfully built, about 5 feet tall, weigh up to 160 pounds and have their cranial capacity between 1,270 to 1,640 cc. The morphology of the pubic bone in the *Neanderthal* suggests that their birth canal was larger than that in the modern man.

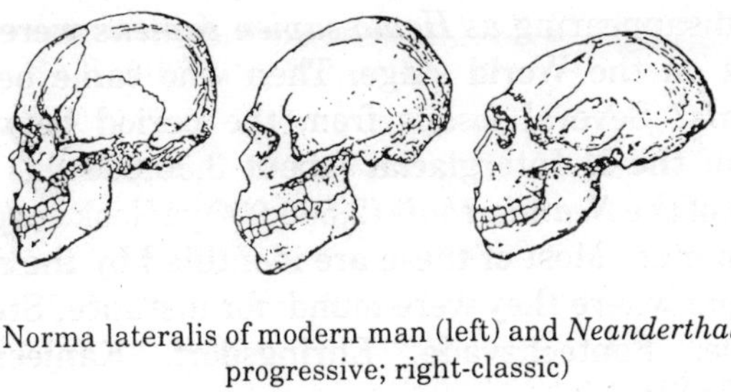

Fig. 6.12. Norma lateralis of modern man (left) and *Neanderthal* men (middle-progressive; right-classic)

The occipital bun (or chignon), previously considered a diagnostic feature for the *Neanderthals*, is now known to be present in the recent humans besides earlier European and African fossils.

The major physical distinctions of the *Neanderthals* compared to the modern humans are:

- They have larger cranial capacities;
- The facial skeletons suggest strong chewing muscles; and
- They have a distinct cranial morphology.

TABLE 6.7
Physical Distinctions between the *Neanderthal* man and the Modern Man (*Homo sapiens*)

Trait	*Homo sapiens*	*Homo sapiens neanderthalensis*
Posture	Upright	Upright
Skull shape	High and narrow	Low and wide
Braincase	High, round dome	Long, flat dome
Cranial capacity	850 - 1,700 cc	1,300 - 1,640 cc
Forehead	Vertical	Sloping
Brow ridges	Small	Heavy
Face	Vertical	Vertical
Chin	Pointed	Absent
Size of teeth	Small	Large
Dental arcade	Arched	arched

Pre-*Neanderthal* Man *(Early Homo sapiens)*

Modern man is believed to have appeared during the IV Glacial of the Pleistocene. *Neanderthals*, i.e., *Homo sapiens neanderthalensis*, existed at least as early as the III Interglacial but began disappearing as *Homo sapien sapiens* were making their debut on the World stage. Then who came before the *Neanderthals*? Several fossils from the period between the beginning of the II Interglacial (about 3,50,000 ya) and the appearance of the *Neanderthals* (2,500,000 ya) look surprisingly like modern man. Most of these are identified by the names of the sites from where they were found: for instance, Steinheim, Swanscombe, Fontechevade, Ehringsdörf, Kanjeera, and Florisbad (Table 6.8).

TABLE 6.8
Early *Homo sapiens*

II Interglacial	Europe	*Steinheim* Man *Swanscombe* Man
III Glacial	Africa	*Kanjera* Man *Florisbad* Man
III Interglacial	Europe	*Fontechévade* (?) Man *Ehringsdörf* Man

Modern man, anatomically, has a much longer history than is implied by the *Neanderthals* and the Upper Pleistocene modern man fossils as above (Table 6.8, 6.9). They appeared earlier than was first thought. Anatomically, modern man was on the scene by the II Interglacial – the date of appearance of the *Steinheim* and the *Swanscombe* Men, the earliest of the pre-*Neanderthal Homo sapiens* fossils. These are the earliest representatives of the *Homo sapiens sapiens*. *Homo sapien neanderthalensis* – the *Neanderthal* population – arose from this line with their characteristically rugged features, which perhaps appeared as an adaptive response to the arctic climate of the IV Glacial. (Remember that the most extreme *Neanderthal* features are found on the fossils of the IV Glacial, while most modern features appear on the earlier, i.e., III Interglacial, *Neanderthals*). The *Neanderthals* than may have been absorbed through interbreeding with the representatives of the new populations of *Homo sapiens* who moved into Europe, perhaps from Asia, during the IV Glacial.

TABLE 6.9
Comparison of Some Important Traits Among Early Man, *Neanderthal* Man and the Upper Pleistocene Man

Trait	*Early Homo sapiens*	*Neanderthal Man*	*Upper Pleistocene Man*
Skull bones	Thick	Thick	Thin
Cranial height	High	Low	High
Occiput shape	Rounded	Angulated	Rounded
Brow ridges	modern man like but protruding	Arched	As in modern man

The *Neanderthal* fossils have been found associated with Mousterian tool tradition mostly characterized by the development of the prepared core technique.

They were big game hunters as revealed by the abundant remains at the Cave site at Shanidar in Iraq and the open air sites like Lebenstadt in Germany: evidence of slaughtered animals, presence of fish and bird remains, etc. They buried their dead following ritualistic behaviour as seen in La Ferrassie, La-Chapelle-aux-Saints, Le-Moustier, and Shanidar Caves, etc.

Relationship between *Neanderthal* Man and Modern Man

Though modern humans and *Neanderthals* are grouped together in the same species, there are differences between their skulls (Fig. 6.13; Table 6.9). Regarding the relationships between them, there exist four probabilities:

- *Neanderthal* man and the Upper Pleistocene *Homo sapiens* are subspecies, i.e., *Homo sapien neanderthalensis* and *Homo sapien sapiens* respectively;
- Anatomically, modern man and the *Neanderthal* man simply represent a population variation as found in the Upper Pleistocene *Homo sapiens*;
- That the *Neanderthal* man is ancestral to the modern man; and
- That the *Neanderthal* man is an evolutionary dead end and modern man evolved from some yet unknown ancestor.

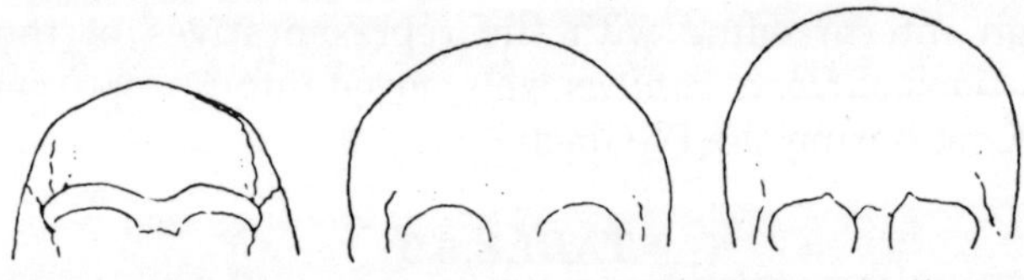

Fig. 6.13. Comparison of the steepening of the forehead in *Homo erectus* (left), *Neanderthal* man (center) and modern man (right)

Subspecies. The extensive fossil finds suggest a gradation from the extreme *Neanderthal* characteristics to some fossils that were very much like modern man. Much of the evidence for the subspecies relationship comes from the two rock shelters on Mount Carmel in Palestine. The female skeleton found in the Tabun Cave has a mixture of the traits both of *Neanderthal* and the modern man, for instance, typically neanderthaloid except for the rounded occiput and the less massive brow ridges (modern man-like). The Es-Skhul Cave yielded a cemetery of ten individuals varying in appearance from quite neanderthaloid to that of the most modern man. These variations in the physical characteristics amongst the *Es-Skhul* fossils (even though there are differences between the modern man and the *Neanderthals*) are large enough to merit division into two separate species; these gradations represent a hybridisation between the *Neanderthals* and the anatomically modern man. Such interbreeding could not occur between different species. Therefore, these two are subspecies, or races of the same species: *Neanderthals* are known as *Homo sapien neanderthalensis* whereas the anatomically modern man is designated as *Homo sapien sapiens*.

Population Variation. This can be supported by the evidence from Mount Carmel in Palestine, which suggests that all the *Es-Skhul* fossils belong to a relatively short span of time. Yet they cover a gradation in physical traits from the extreme neanderthaloid to the relatively modern human types (Fig. 6.14). This clearly indicates that these fossil remains cover the entire range of variation in physical types of that particular period.

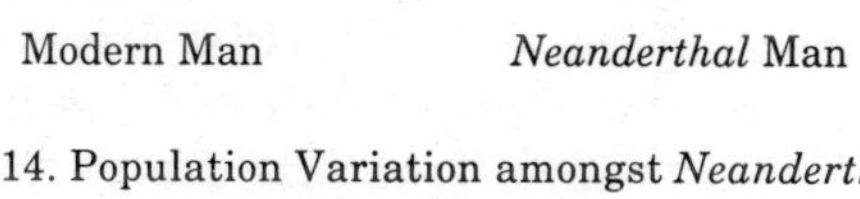

Fig. 6.14. Population Variation amongst *Neanderthal* man and modern man

This interpretation agrees with the genetical concept of a species as a population, or group of populations, of actually interbreeding organisms that are reproductively isolated from other such groups.

Ancestral. These fossils date from an earlier time than those of the anatomically modern man. The physical form of *Neanderthal* man is a logical halfway point between the fossil forms that came before it and the anatomical modern man. Gradations in physical types in the cemetery at Es-Skhul provide evidence of a transition from the *Neanderthal* man to the modern man in a straight line – evolutionary scheme (Fig. 6.15).

Homo sapien sapiens

↑

Homo sapiens

↑

Homo sapien neanderthalensis

Fig. 6.15. Transition from *Neanderthal* man to modern man

The 10,000 years overlap between about 50,000 to 40,000 years BP, during which both the *Neanderthal* and the anatomically modern man's fossils are found, is also interpreted as the evidence of a straight line evolutionary change from the *Neanderthals* to modern man, with the 10,000 years as the span of evolutionary change from one to the other. This probability, however, has a limitation namely that it presumes a very rapid rate of gene mutation or gene variability – rate of change, which is not likely to occur.

Separate Lines. The basis for this probability is the relatively abrupt transition from the *Neanderthal* man to the modern man in the fossil records providing evidence that modern man moved into Europe from somewhere else, perhaps from Asia, and killed off the less capable *Neanderthals* in a primitive genocidal conflict. A theory that cannot be challenged until all fossil men sites are excavated! The major opposition to this theory is that the Mount Carmel fossils represent a physical transition between the *Neanderthals* and the modern man is based on cultural evidence. True that one prehistoric culture, the Mousterian industry, is found only in conjunction with the *Neanderthal* remains whereas the innovations in the tool making industry appear with the first fossil remains of *Homo sapiens*. But there exists no sufficient reason to equate new tool types with absolutely new forms of 'man'! New tool techniques are generally based on the old techniques, and many of the old tool types still appear simultaneously with the new tools. Even if the anatomically 'modern' man did migrate into the

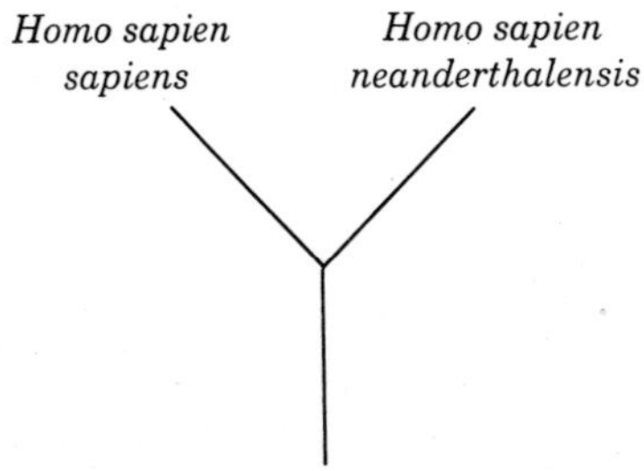

Fig. 6.16. Another hypothesis of transition from the ancestral line

Neandarthal territory of Europe, it is most likely that they simply interbred with the indigenous population, gradually mixing both their cultures and their genes to yield 'man' as he is today (Table 6.10).

TABLE 6.10

Comparison between the Early *Neanderthals* and the Late *Neanderthals*

Traits	*Early Neanderthals*	*Late Neanderthals*
Body structure	Light	Robust
Brow ridges	Light	Heavy
Cheeks	Slightly hollowed	Not hollowed
Occipital bun	Sometimes absent	Present

7

Modern Men

Cro-Magnon

Cro-Magnon man is the name given to a well-known population of prehistoric humans who lived in late Pleistocene times. Skeletal remains of this type were first discovered in 1868 when some road builders, working near the Vezere River in Dordogne, Southern France, accidentally opened a prehistoric rock shelter. Besides human bones, stone tools and ancient hearths were also found at the Cro-Magnon site, which is only one of many prehistoric caves and rock shelters in the area.

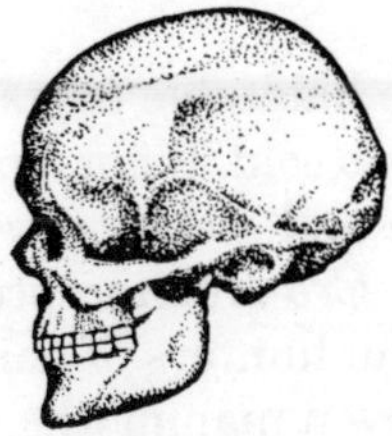

Fig. 7.1. Lateral View of *Cro-Magnon* Skull

Cro-Magnon is the name of a rock shelter near Les Eyzies-de-Tayac, Dordogne (France), where several prehistoric skeletons were found in 1868. Excavations revealed the existence of five archaeological layers covered with ash (Lartet, 1868).

Although it appears, that at the time of discovery, the remains of more than 10 individuals existed at Cro-Magnon, only fragments of five were preserved and studied. Among these was the cranium and mandible of a male about 50 years old. This specimen is known as the "Old Man of Cro-Magnon" (Fig. 7.1). The name *Cro-Magnon* is now generally referred to all such anatomically modern *Homo sapiens* dating from the Upper Palaeolithic period (c. 35,000 to 10,000 years ago) found in Europe and the Middle East, from their rather sudden appearance about 40,000-30,000 years ago. In contrast to the early *Neanderthals*, the *Cro-Magnons*, classified as *Homo sapiens sapiens*, possessed such modern characteristics as higher, more vertical foreheads – suggesting a modern configuration of the frontal lobes of the brain – reduced brow ridges, smaller faces and teeth, and a chin.

Cro-Magnon people, the first examples of prehistoric art, appeared in the Old World. Their remarkable cave paintings and engravings, primarily of animals but also of non-representational subjects, are visible on the walls of many caves in Spain and southern France, including the famous sites of Lascaux and Altamira. These artistic efforts, along with the finely made tools, testify to the high level of cultural sophistication reached by *Cro-Magnon* man. Although there are many samples of crude *Neanderthal* tools, and also evidence of rituals, these early "humans" left no evidence of any pictorial or sculptured art.

Hunting Techniques

The climate in the habitable parts of Europe during *Cro-Magnon* times was cool to cold. Plants and animals of the types associated with tundra and steppe environments were usual. Bone remains found at *Cro-Magnon* occupation sites indicate that they were successful hunters of animals such as reindeer, bison, wild horse, and even mammoths.

Place in Human Evolution

The question of the relation of *Cro-Magnons* to the earliest forms of *Homo sapiens* is still unclear. It does appear, however, that *Cro-Magnon* (*Homo sapiens sapiens*) and *Neanderthals* (*Homo sapiens neanderthalensis*) are close than was once believed. It was long thought that certain *Cro-Magnon* traits could be seen in them

Culture

The relations among *Homo sapiens sapiens*, and particularly *Cro-Magnon* peoples and the various Upper Palaeolithic cultures (e.g., Chatelperronian, Aurignacian, and

Gravettia, which are classified on the basis of stone and bone tools) are relatively distinct; although in 1979 itself, *Neanderthal* fossils were found in Chatelperronian.

Eyzies-de-Tayac Caves. Series of prehistoric rock dwellings have been located downstream from Lascaux Grotto and near the town of Les Eyzies-de-Tayac in Dordogne (South Western France). The caves include some of the most significant archaeological finds of the Upper Palaeolithic (from about 40,000 to 10,000 years ago).

The Tautavel Man. The skull of the young adult *Tautavel* "Man" (450,000 years ago)#, which was discovered in 1971, from the Arago Cave$ has a flat and receding forehead and a well-developed eyebrow region. The face is big with rectangular orbits. The height of the *Tautavel* Man is 1.65 m and has sturdier bones than those of present day humans. The cranial capacity is 1150 cc.

Swanscombe Man

A more widely known example of early *Homo sapiens* is derived from the Thames Middle Gravels at Swanscombe in North Kent, England; these gravels are attributed to the Mindel-Riss interglacial stage, 400,000 to 200,000 years ago. At this site, which is well known for its Palaeolithic tools, two parts of a human skull (Fig. 7.2) were found in 1935 and 1936. About 20

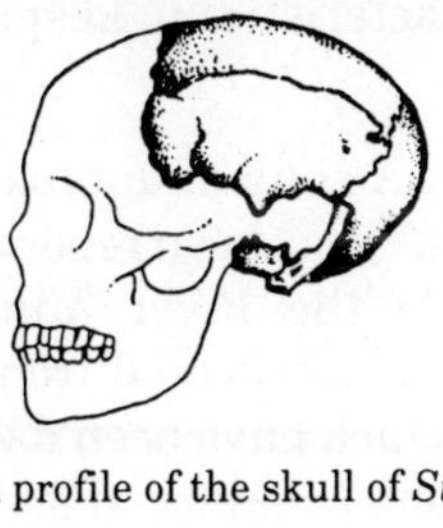

Fig. 7.2. Lateral profile of the skull of *Swanscombe* man

years later, a third piece was recovered that fits with the other two pieces to form the back half of the cranial vault. All these

Age estimated using the direct, non-destructive dating method.

$ The Arago Cave, located in Tautavel Town, is cut into urgonian limestone and projects several hundred metres over the Valley crossed by the Verdouble.

three bones are almost complete, and they probably belong to a young adult. Generally speaking, the bones are modern and rather thick; the skull is broad at the back. The occipital bone does not have a heavy torus or "chignon" – the characteristic of the *Neanderthals*. The occipital ridge is modest in its development and a little more prominent at its ends rather than in the middle. The bone is rounded, both above and below the ridge, in contrast to that in the *Vertesszollos* occipital bone. The position of the foramen magnum does not differ from that in the *Homo sapiens*. It is argued that on the basis of the measurements of the bones, the *Swanscombe* skull could not be distinguished from that of the *Homo sapiens*. Despite this, several workers have been less convinced by the sapient features of *Swanscombe* man and have emphasized the *Neanderthal* features of the bones. A more recent reappraisal of the *Swanscombe* material, making use of the multivariate statistics, suggests that it is necessary to emphasize the interrelatedness of all the early forms of *Homo* and that it is no longer possible to maintain specific status for each form. It is preferred, rather, to regard them all as a spectrum of varieties within one species. It is further suggested that *Swanscombe* should be regarded as belonging to a "Neanderthaloid Intermediate" group that contains *Ehringsdörf*, *Skhul V*, *Krapina,* and *Steinheim* specimens. As with the *Vertesszollos*, there is a mixture of features – at this somewhat later date in Europe, which combines sapient and *Neanderthal* characteristics in keeping with the concept of mosaic evolution.

Steinheim Man. A skull of approximately the same age, as the *Swanscombe* skull, has been reported from the Steinheim site in 1933 alongside the River Murr, near Stuttgart in Germany. This skull was recovered from a gravel pit cut into Pleistocene deposits, which have been dated to the Mindel-Riss Interglacial stage (Second Interglacial Period). Though this skull is more complete than the others mentioned above, but is distorted, perhaps due to the pressure of the deposits. It consists of the cranium and the right side of the face of a young adult; much of the base of the skull is missing. However, larger part of the palate, and some teeth including a PM and all the molars of both the sides are present. The face is straight, with slightly protruding upper jaw. The vault of the skull appears to be long and narrow but fairly well rounded in profile. The skull has only moderate frontal flattening. The rounded occipital region is marked by a very low occipital ridge and lacks the 'occipital bun' characteristic of the *Neanderthal*. The frontal region has a pronounced and moderately heavy but divided brow-ridge. The cranial capacity is estimated to be within 1,070 and 1,175 cc; the distortion, however, precludes accurate measurement.

The teeth are of particular importance because they are perhaps the earliest sapient teeth known today. The crown of the PM is symmetrical in shape but has a large outer and a smaller inner cusp. The molars decrease in size from front to back; the third molar is markedly smaller than the other two. All the teeth have slight degree of taurodontism (i.e., enlarged pulp cavities). However, it does not deviate from the normal range of variation for these traits in modern man. *Steinheim* is usually classified as an early subspecies of *Homo sapiens*, called *Homo sapiens heidelbergensis* (Fig. 7.3).

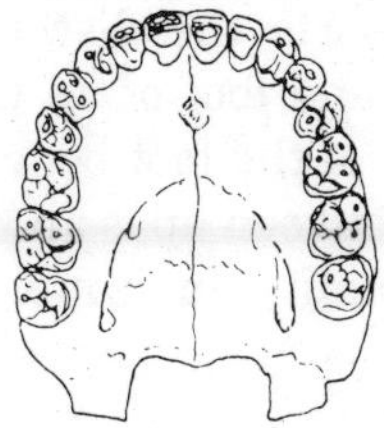

Fig. 7.3. *Homo sapiens heidelbergensis*

This skull has a mixture of primitive and advanced characteristics. Its main resemblance is with the *Swanscombe* Man (Fig. 7.2). The contour of the occipital region and the rounded vault display advanced sapient traits, whereas the broad nasal opening and the brow ridges resemble the *Neanderthal*. The teeth are small and sapient in, both form and molar size; the marked reduction of the size of the third molars is distinctly a modern characteristic. The enlarged pulp cavities are often regarded as primitive for they are widely seen in both *Homo erectus* and *Neanderthal*. However, this feature is still frequently observed in the teeth of modern humans.

Earlier opinions suggest that *Steinheim* man represented

an intermediate stage between *Homo erectus* and the later forms of *Homo*, thus, being ancestral to both the *Neanderthal* and modern man. Yet, others do not accept the *Steinheim* man as being ancestral to the classic *Neanderthal* and modern man suggesting instead that it represents a stage on a separate but parallel line leading to modern man. Both the views appear to be contrary to the findings of modern population genetics; the occurrence of a combination of characters, which is becoming apparent, follows the general trend of the cranial enlargements and rounding, accompanied by a reduction in the facial and dental characteristics.

Discovered by gravel pit workers in 1907 near Heidelberg in Germany. It's estimated age is between 400,000 and 700,000 years. This find consisted of a lower jaw with a receding chin and all its teeth. The jaw is extremely large and robust, like that of *Homo erectus*, but the teeth are at the small end of the erectus range. It is therefore identified as *erectus* on the basis of its age, but could be an archaic *sapiens*.

Clarke (1997) reports the discovery of the articulating bones of the lower leg and foot of an upright walking ancient human ancestor. Some of the foot bones of these ancestors are intermediate in form between apes and humans. This suggests some degree of arboreality in combination with terrestrial bipedality.

8

Human Evolution

Human beings, extant and extinct, comprise the zoological family Hominidae. The single living human species, *Homo sapiens*, is one of the species of the order *Primates*, which in turn is one of the 20 orders constituting the vertebrate class *Mammalia*. Among the past and present diversity of primates, hominids have long been recognized as having the closest resemblances, and hence affinities, to the African great apes (pongids). Thus, in 1863, the British biologist T.H. Huxley noted in *The Evidence as to Man's Place in Nature* that "whatever system of organs be studied . . . the structural differences which separate Man from the Gorilla and the Chimpanzee are not so great as those which separate the Gorilla from the lower apes [monkeys]." Various methods for the comparative evaluation of genetic characters have both repeatedly confirmed and measured in some detail the very close proximity of the extant African apes and modern *Homo sapiens.* All such findings are congruent with a common origin of apes and Hominidae, within the African continent, which took place some five to six million years ago.

Three major areas are generally recognized within the subject of human evolution: primatology, which has as its major focus on the biological and behavioural aspects of nonhuman primates; human palaeontology, which is concerned with the recovery, description, and evaluation of the fossil evidence for hominid evolution; and palaeoanthropology, which encompasses interrelated investigations into the biological and behavioural evolution of Hominidae. In addition, five major areas of research can be identified in human evolutionary studies: the origins of Hominidae, adaptation and diversification of the genus *Australopithecus*, the origins of the genus *Homo*, the emergence of *Homo erectus* and subsequent hominid occupation of Eurasia, and the origins and dispersals of premodern and modern *Homo sapiens*.

Investigations of hominid origins are variously concerned with diverse comparative studies of extant higher primates and humans, as well as the search for ancestors in the fossil record. Pongids and hominids show a diversity of contrasting adaptations that evidently reflect their evolutionary divergences and which thus require explanation. Moreover, although markedly different from the Asian pongid (orangutan), the African pongids (gorilla and chimpanzee) differ from one another both structurally and behaviourally. The roots of Hominidae have been traced to at least four million years ago, and possibly to some five million years ago. The rarity and fragmentary condition of the few oldest known specimens, however, do not reveal critical aspects of hominid adaptation, such as modifications in trunk and lower-limb structure. Hence, the details of hominid origins remain unknown and the subject of lively debate and substantial speculation. The ancestral stock of extant African apes and of hominids also remains unknown, in large part, a reflection of the paucity of fossil-bearing localities in the five- to 10-million-year time span. In the absence of a fossil record, structural and other adaptations have been projected back as an ancestral condition from living descendant species; but this is a very risky procedure that dismisses morphological transformation and adaptation and assumes stasis without complementary confirmation.

The Antiquity of Homo Sapiens

It is a curious fact that, although evidence for the evolution of man is extensive, direct fossil evidence of the earliest members of the species *Homo sapiens* is relatively scarce. The species *Homo sapiens* (of which the modern human races comprise a number of different geographic varieties) may be defined in terms of the anatomical characters shared by its members. The definition for the prehistoric representatives of the species must be limited to skeletal characteristics and includes such features as:

- a mean cranial capacity of about 1,350 cubic centimeters (82 cubic inches);

- an approximately vertical forehead;
- a rounded occipital part of the skull with a relatively small area for the attachment of the neck musculature;
- jaws and small sized teeth; small canine teeth of spatulate form;
- the presence of a pointed or projecting chin; and
- limb bones adapted to a fully erect posture and gait.

Any skeletal remains that conform to this pattern to an extent that precludes classification in other groups of higher primates must be assumed to belong to the anatomically modern *Homo sapiens*. In the past, there was a tendency to create entirely new species of *Homo* on the basis of fragments of prehistoric human skeletons, even though the remains showed no significant differences from modern man. This tendency was prompted by the supposed antiquity of the remains or by a failure to realize how variable some features are even in modern man. The species of the genus *Homo* that immediately preceded *Homo sapiens* was *Homo erectus*, and it is most likely that sapient humans (*Homo sapiens*) evolved from *Homo erectus*.

Structure of Homo Sapiens. The bodily structure of the fossil remains of early *Homo sapiens* are known from sites in Africa, the Middle East, and Europe, but later examples come from a wide range of sites in the Old World as a whole. It is of particular interest to look at the skulls of some of the early specimens, for it is in their functional morphology that the combination of features that results in attribution to *Homo sapiens* are often found (Fig. 8.1).

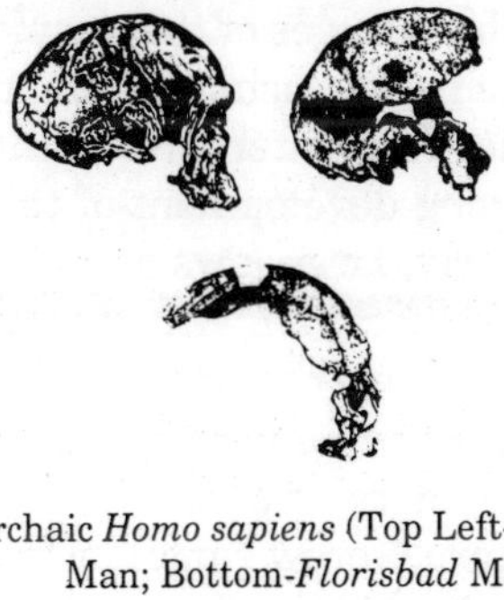

Fig. 8.1. Skulls of Archaic *Homo sapiens* (Top Left-*Jebel* Man; Top Right-*Sale* Man; Bottom-*Florisbad* Man)

Evolution of the Human Skull. The human skull is composed of both cranial and facial portions. The cranium (also known as neurocranium) consists of the skull vault and base, while the facial skeleton (also known as splanchocranium) consists of the region of the eye sockets, nose, cheekbones, upper jaw (orbital and maxillary region), and the region of the lower jaw (mandibular region). During the evolution of the hominid skull from its ape-like precursors, there are a number of general trends that can readily be discerned. The principal trends are the gradual increase in brain size (as measured by cranial capacity), the rounding of the cranial vault, and the gradual reduction of the size of the whole masticatory complex, including both the upper and the lower jaws and the teeth. These trends lead to an overall change in skull shape and proportions, so that, while the vault expands, the "muzzle" tends to retract from a protruding (prognathic) form to a straighter-faced (orthognathic) appearance. At the same time, the whole skull tends to become lighter and more delicate in its structures. If this process is examined a little more closely, it can be at least partly explained in mechanical terms. A comparison of the skull of *Homo erectus* and a modern *Homo sapiens* skull illustrates the point.

The skull of *Homo erectus* has a low skull vault, a sharply receding forehead, a lower cranial capacity (averaging less than 1,000 cc), a large face with big teeth and jaws, and a jutting occipital region (the occipital bun). Muscles in the occipital region relate principally to the balance of the head on the vertebral column (spine); those of the temporal region (at the sides of the head) relate to the function of the jaw. In a skull with a large face and a heavy jaw, the muscles of the occipital region, which balance the head on the neck (nuchal musculature), are strong (well-developed) in order to counterbalance the heavy face and jaw. This results in strong development of the occipital ridges in *Homo erectus*. Similarly, large jaws and teeth demand well-developed muscles of mastication with strong attachments to the cranium. In order to counterbalance the considerable forces produced by these masticatory muscles when the teeth occlude (meet) during chewing, there is a stout maxilla and a heavy supraorbital ridge, or torus (brow ridge) in *Homo erectus*. This torus extends the whole width of the cranium (even more necessary because *Homo erectus* had a lower cranial capacity and therefore lacked a well-developed forehead). Such skulls are found in the fossil remains of *Homo erectus* known from both China and Java.

By comparison, the skulls of *Homo sapiens* show an expanded cranial vault with a high maximum breadth and a well-developed vertical forehead (resulting from expansion of the frontal region of the brain) so that the supraorbital ridge is, as it were, "overgrown" and buried. At the same time, the face is shortened by a reduction in size of the jaws, which also have smaller teeth (Fig. 8.2). The masticatory muscles, therefore, need not be so large; and the forces that they produce are less, so that the need to neutralize them through the face is reduced. Thus, the facial structure is more delicate, and there is no need for a heavy supraorbital ridge. Finally, the lighter face does not require powerful muscles in the occipital region to balance the skull; as a result, the neck muscles are reduced in size and the area for their attachment to the skull is more underneath the back of the head.

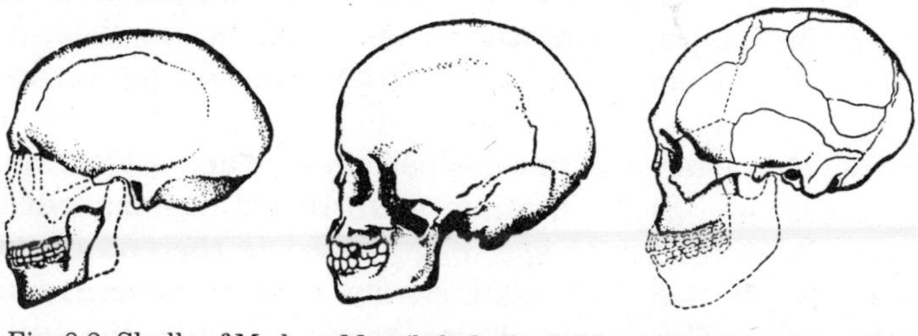

Fig. 8.2. Skulls of Modern Men (left-*Galley* Hill; middle-*Grimaldi*; right-*Wadjak*)

The decrease in the size of the teeth of *Homo sapiens* tends to leave the nose and point of the jaw as prominent features of the face. Thus, the presence of a mental protuberance (chin) is an obvious character of the jaw of *Homo sapiens* and provides external support for the symphysial region (the point at which the two halves of the jaw are joined together).

The oldest definitely known hominids are attributed to the extinct genus *Australopithecus*. The genus speciated substantially, producing distinct and, in some cases, possibly convergent lineages. At least four species (*afarensis, africanus, robustus, boisei*) are commonly accepted, and two more (*aethiopicus, crassidens*) are recognized by some workers on morphological grounds. All species of the genus originated in the Pliocene epoch (5.3 to 1.6 million years ago), and the genus apparently became extinct in the Early Pleistocene (about 1.6 million to 900,000 years ago); its distribution is unknown outside the African continent. The oldest and most primitive species is *Australopithecus afarensis*, and most workers believe it to be ancestral to succedent species. Although they exhibited some fundamental hominid adaptations (bipedalism, reduction of anterior dentition, exploitation of non-forested habitats), most or all australopithecine species remained primitive in terms of growth and maturation, brain size and proportions, dietary adjustments, and complexity of cultural behaviour. *Homo* coexisted with the later, so-called *robust*, australopithecines – *robustus* (and possibly *crassidens*) in southern Africa and *boisei* in East Africa – although the adaptations enabling such coexistence are scarcely understood, and it is generally thought that an australopithecine species was ancestral to *Homo*.

The recognition and suitable definition of the genus *Homo* and its initial representatives has been a persistently troublesome problem. There have been no formal diagnoses, and the few characterizations offered suffer from both lack of definitive characteristics and inclusiveness. The problem has been exacerbated as the hominid fossil record has expanded, particularly in respect to specimens dated to the end of the Pliocene epoch that lack distinctively *Australopithecus*-like characteristics. The first such specimens, found in the early 1960s in the Olduvai Gorge of Tanzania, were designated *Homo habilis*. Further remains of both comparable and greater age were subsequently recovered from northern Kenya and south-western Ethiopia. Although a single initial *Homo species* (*Homo habilis*) was originally proposed, this perspective has been criticized by some workers as simplistic because of the substantial variability of the fossil finds; accordingly, it is entirely possible that two contemporaneous and even sympatric species may have existed in the Late Pliocene. Coincident with the appearance and subsequent presence of such hominid(s) are various traces of associated culturally patterned behaviours. These include evidence of natural but transported and accumulated stones, flaked-stone artifacts, and occasional associated mammal (and other) skeletal parts, all of which indicate the exploitation and utilization of animal resources;

the repeated utilization and occupation of particular locales; and the expanded employment of natural resources in conjunction with technological capabilities and requirements. Such biological and behavioural adaptations are believed by many workers to reflect major transformations and reorganizations in hominid phylogeny, perhaps consequent upon the initial appearance of genus *Homo*.

The fossil record in sub-Saharan Africa affords evidence of the appearance of another, more derived (i.e., more evolved) species of *Homo* – *Homo erectus* – at the beginning of the Pleistocene epoch. At several localities in East and Southern Africa, the species occurs sympatrically with the "robust" australopithecines. Less ancient occurrences are also known from north-western Africa. The initial occupation of Eurasia by hominids appears to postdate such an antiquity, and it is generally inferred that the first Eurasian hominids were dispersals from an African source, perhaps between 1.5 and one million years ago. *Homo erectus* was first and, for a long time, best known from fossil finds in Southeast and East Asia. The fossil occurrences there range in age from approximately 1.6 million to 250,000 years. Although initial hominid occupation in Europe was probably at least as early, no human skeletal remains are known from the most ancient times, and those that have been found – dated to between 500,000 and 300,000 years ago – do not represent *Homo erectus* but rather a form of *Homo sapiens* that has been labeled "archaic." The initial and subsequent penetration of hominids from the lower to middle latitudes occurred as the amplitude and intensity of glacial-interglacial climatic cycles was increasing. The extent to which there were attendant, and perhaps correlative, changes in human biology and in behaviour have remained the subjects of substantial research and much controversy.

Traditionally, the tendency among students of hominid evolution was to attribute premodern human fossil finds to one or more extinct species, and sometimes even distinct genera, but as efforts increasingly have been directed at seeking congruence between developments in evolutionary biology and in the state of the hominid fossil record, substantial revisions in the classification of Hominidae have emerged. A variety of premodern human finds of both late Middle Pleistocene and early Late Pleistocene ages came to be subsumed within the species *Homo sapiens* and were only further distinguished below the species level. This varied and increasingly large sample of post-*erectus* hominids came to be regarded as archaic *Homo sapiens*, as distinguished from anatomically modern humans. This is largely a consequence of the mosaic of morphological features represented in the substantial variations among such materials, which exhibit primitive versus advanced, or derived, features.

If early African *Homo erectus* constituted the source of subsequent hominids, then it would appear that evolution proceeded quite differently in major geographic areas. For example, *Homo erectus* is characteristic of the Asian Middle Pleistocene, where it is also long persistent and distinguished by its own group of singular, derived features. In Africa, derivatives of early *Homo erectus* are known, as are some transitional examples linking *Homo erectus* to archaic *Homo sapiens*. Skeletal parts of the earliest hominid occupants of western Eurasia are not known. The first examples from this area occur well into the Middle Pleistocene, and there is a range of variation in the specimens from strongly and partially *erectus*-like to incipiently *Neanderthal*-like, passing ultimately into the well known and widely distributed *Neanderthal* people. The *Neanderthals* were for many years treated as a distinct species (*Homo neanderthalensis*), but they were subsequently subsumed as an archaic subspecies of *Homo sapiens*. An increasingly substantial body of evidence has been accumulated, however, which suggests that a return to the older position is probably warranted. In Western Europe, at least, there is increasing evidence of the contemporaneity of the last *Neanderthal* peoples with those early modern populations that have come to be known as *Cro-Magnons*. For these reasons, and in order to recognize and express differing degrees of derivation, further taxonomic evaluation and distinction of these archaic *Homo sapiens* specimens is required, which will doubtless include the recognition of additional subspecies of both the *Homo erectus* and *Homo neanderthalensis* groups in Western Eurasia and also corresponding taxonomic reassessment of various African Middle Pleistocene samples. In East Asia the existence and morphology of archaic *Homo sapiens* has been well established, but the extent to which this form was a contemporaneous or a succedent replacement for the late *Homo erectus* populations has not been firmly resolved.

The roots of anatomically modern humans have long been a puzzle to students of human evolution and hence the source of much speculation and debate. Nonetheless, several developments have caused renewed interest in the problem from different perspectives. First, there has been the recognition of substantially greater relative and absolute ages for the archaeological industries of the African Middle Stone Age (about 200,000 to 40,000 years ago) and, correspondingly, some associated modern-like human skeletal parts. A series of such sub-Saharan occurrences has been identified within earlier and later segments of this time span. Second, there has been the recognition and broad acceptance of early modern human (often called *Cro-Magnoid*) populations in western Asia that were distinct from, and considerably older than (i.e., 90,000-100,000 years ago), the known *Neanderthals* from that area; and also were much older than the European *Cro-Magnon* peoples who were widespread in Europe by some 30,000-35,000 years ago. Third, there has been the increasing availability of comparative genetic data on degrees of affinities of modern human populations. The biochemical systems of Asian and European populations appear to be more similar to each other than those of either group are to African populations. Thus, Asians and Europeans may have shared a common ancestry some 40,000 years ago and a common ancestry with African populations almost three times as long ago. Moreover, investigations of human mitochondrial DNA reveal two facts: that the variation among modern human populations is small compared, for example, with that between apes and monkeys. This points to the recency of human origin; and that there is a distinction between African and other human mitochondrial DNA types, suggesting the substantial antiquity of the African peoples and the relative recency of other human populations.

9

Human Origins

Few issues in anthropology have generated more debate, over the past few years, than the biological and behavioural origins of modern humans. This is evident from the number of important reviews dealing primarily with the biological aspects of the emergence of modern human populations, especially the phylogenetic relationships between anatomically *archaic* and anatomically *modern* populations in different regions of the world. Human understanding of the evolution of the hominoids and the hominids has, of lately evolved into a multidisciplinary pursuit involving geologists, ecologists, molecular biologists, and zoologists etc. Varying views on the hominoid and hominid evolution have been appearing during the last two decades or so. Today, the anthropologists are debating as to where did the modern humans originate? Some favour Africa, others Asia, and still others Europe as the place of origin. Still others believe that this transformation took place, simultaneously, in many different areas. The issue of "population continuity" versus "population replacement" has been the important aspect of the debate over the origins of modern human populations throughout the greater part of last century. Broadly speaking, one might say that the opinion derived from the fossil skeletal material has been more or less equally divided. For instance, with a succession of workers arguing for varying degrees of population dispersal and population replacement over the period of the archaic/modern-human transition and others arguing for an essentially gradual, continuous process of evolutionary change. The recent developments in the field of molecular genetics are potentially crucial to this debate.

Ape-like higher primates had a widespread Eurasiatic distribution (up until five to ten mya, by which time such creatures had disappeared from increasingly temperate Europe); presumably, they were also common in some parts of Africa where fossiliferrous beds of that age are now singularly rare (Howell, 1964). The rare and fortunate occurrence of skeletal parts, for instance, limb bones, indicate that some distinguishing features of modern apes were later evolutionary "specializations" rather than the primitive ancestral conditions. Several specimens of jaws and teeth, from regions as widely apart as Northern India and Eastern Africa, and about 10 to 14 my in age, reveal some hominid similarities. The hominoid fossil record has until recently been both fragmentary and poorly dated. Though the living large hominoids do not vary widely, the Middle and Late Miocene hominoids show great diversity. As a consequence, the fossil evidence of the past hominoids is quite heterogeneous. The recent finds, however, are providing vital clues necessary for the reconstruction of their branching. A variety of factors have no doubt contributed to this, including not only the discovery of a range of important human skeletal remains but also radical improvements in the techniques for dating the finds and major advances (both theoretical and methodological) in the study of archaeological records of human behaviour. The most significant development, however, has been the recent research into DNA and unrelated aspects of human genetics, which is opening up a whole new dimension in our understanding of the origins and evolutionary relationships of modern human populations (Mellars, 1989). Biomolecular studies too suggest a realignment of the evolutionary course of the hominoids. The recent discoveries of early hominid antiquities from Tibet and Northwest should arouse fresh interest in our understanding of the human origins. What happened to these early hominoids in this region?

Anthropologists and archaeologists have conducted extensive excavations in Africa, Europe and Asia trying to find fossil evidence that will answer the most fundamental questions of our existence: When, where and how did the modern humans arise? Yet, despite more than a century of digging, the fossil record remains too sparse. We are in a position of being able to perceive many of the essential patterns in human cultural and biological development but far less able to offer coherent and convincing explanations for the patterns we observe. Having such a vast range of geographical and heavily researched areas, databases, and theoretical perspectives under consideration, it is hard to do justice to these long standing discussions at this juncture. It is perhaps inevitable that much of the critical debate today centres on the rapidly accumulating evidence for a primary origin of the anatomically modern human forms in the Asian zone. The evidence from further east in Asia remains poorly documented, poorly dated, and – at least as regards the archaeological material – rather poorly described (Mellars, 1988).

As early as 1868, the coexistence of extinct animals such as mammoths with anatomically modern, but very robust, humans was documented at the site of Cro-Magnon, in Southern France.

Anatomically modern humans are distinguished from their predecessors by their relatively "gracile" (less robust or less muscular) skeletons and smaller teeth. Males, in particular, became smaller and overlapped the female size range to a greater extent than previously. Although brain size did not increase in moderns from the preceding "archaic" stage, the braincase itself became taller, less elongated from front-to-back, and more sharply flexed at its base, where it joins the face. In short, the face became almost completely situated under the braincase, rather than sticking out in front of it as in earlier human ancestors and other primates. Smaller teeth also left the chin sticking out in front, and reduced the need for heavy brow ridges to take up some of the stress of chewing. Archaic *Homo sapiens*, with modern-size brains but big brow ridges, large faces, and large teeth, occupied Europe, Asia and Africa before the appearance of modern *Homo sapiens*. The term *Neanderthal* refers in some theories to one relatively isolated cold-adapted population of these "archaics". In other theories, *Neanderthal* refers to all later "archaics", ca. 130,000 to 40,000 B.P.

During the last few decades, two basic views have vied for acceptance by the scientific community. The "candelabra" story recognizes only one major branching of the human line. After the initial dispersal of humans to the three major Old World Continents, beginning as early as 1.1 mya with the species *Homo erectus*, the populations of each region evolved in parallel fashion into modern humans. Some migration or gene flow between the regions assured that new characteristics appearing in one region would eventually spread to all. According to this viewpoint, most of the immediate ancestors of the modern humans of Africa are found in Africa, while the immediate ancestors of Chinese are found in China and so on. The immediate ancestors of Europeans, according to this theory, are the predecessors, namely the *Neanderthals*, on that continent. The "multi-regional evolution" or the "candelabra" theory, as it is referred, allows more migrations from region to region than earlier theories, i.e., it suggests a lot of gene flow between regions, and implies very deep roots to regional populations.

On the other hand, the "hatrack" theory suggests a single main stem or center pole leading to modern humans, with branches, at intervals, through time representing evolutionary dead ends. The *Neanderthals* of Europe are one such dead end and the "*Peking* Man" or *Homo erectus* of East Asia is another dead end according to this theory. Till recently, the central main pole was always given a European or Near Eastern identity, through such fossils as *Swanscombe* (dating to a period just before the earliest *Neanderthals*), or the *Skhul* fossils from Israel. Some attributed this central role of Europe in human evolution to the influence of a colder climate, a limited growing season, more reliance on both hunting and food storage, all of which would have promoted intelligence and growth of the brain. According to the "hatrack" theory, the central stem is African, and all the earlier fossils of other continents constitute the dead ends of human evolution. All anatomically modern humans are derived from recent African ancestors: this modern theory is called as the "out-of-Africa" hypothesis.

How can two different views continue to co-exist? Why does not the data exclusively support one or the other hypothesis? New dates, new data and new DNA studies throw newer insights over modern human origins.

Dating

The shift to modern humans was virtually complete throughout Europe, Asia, Africa and even Australia around 35,000 years ago. The most accurate dating technique for the later periods of archaeology, radiocarbon, gives good results back to about 35,000 years, but not much older. Some dates, however, of 38,000 to 40,000 are acceptable, but dates in the range of 40,000 or older are certainly dubious. Interestingly, most of the story of modern human origins lies beyond 40,000 years ago. Until recently, there were no reliable ways to determine the age of anything between 40,000 to 200,000 years ago.

However, a range of new techniques has recently come into general use for exactly the period when modern humans must have emerged, f.i. between 200,000 and 40,000 years ago. These techniques include:

- measuring the accumulation of "radiation damage" from soil radiation in buried crystalline materials such as flints or quart sands (thermoluminescence);
- measuring the decay of uranium which soaks/into buried bones and teeth from groundwater (uranium series) or radiation damage in crystals of tooth enamel (electron spin resonance); and
- studying the decay of the proteins encapsulated in hard tissues of fossil animals such as mollusc shells, bones, teeth, and ostrich egg shells (amino-acid racemization).

Unlike radiocarbon, none of these techniques is entirely independent of the burial environment. Thermoluminescence and electron spin resonance dates can be thrown off by inaccurate measurement of the soil radiation or by heating or re-exposure of the sample before the archaeologist finds it. Protein decay rates, on the other hand, are dependent on temperature, which is difficult to estimate for 40,000 to 200,000 years ago. And the uranium, which soaks into bones and teeth, can also wash out again. Therefore, using two different techniques to date the same site can help avoid these problems, at least when the two sets of results agree.

The European dates did not change quite as much, because the ebb and flow of ice ages had provided a chronology that tied most of the sites together, even in the absence of exact numbers.

As soon as the chronology of Africa was based on its own internal sequence of dates, comparative faunal extinctions, and climate changes, it became obvious that the earliest fossils in Africa with "chins" and small teeth were much older than the *Cro-Magnons* of Europe. It has also been suggested that several of the most important early African sites with modern humans date to as much as 105,000 years ago or older. Modern human teeth at Mumba shelter in Tanzania were dated to ca. 130,000 years by uranium series.

New dates for Peking Man sites and other sites from China and Java suggest that East Asia was occupied exclusively by the more primitive species *Homo erectus* until about 300,000 years ago. (That the earliest modern humans were African seems quite well-established, although very few sites have been dated thus far).

The fossil record is far too patchy to say how *Homo habilis* (handy man) and other members of its genus came to stand out among the *Homo* line and eventually led to modern humans. *Homo erectus* was also the first hominid to emigrate from Africa, at least 1.8 million years ago, spreading all the way to China and Indonesia. Then, at some point – for reasons still mysterious – the lineage diverged, with one branch leading to the *Neanderthals* and another to the modern humans. Exactly when and how it happened is unclear. The oldest *Neanderthal* fossils date only to 200,000 B.P., and the oldest *Homo sapiens* to about 100,000. But some recent discoveries may help answer those questions. For instance, a one million year old cranium from Buia, Eritrea has characteristics of both *Homo erectus* and *Homo sapiens* (Abbate *et al.,* 1998). The "spectacular" partial cranium of the same age from Ethiopia should help as well when it is formally described.

The principal effects of the new dates in Europe have been twofold: to demonstrate the great antiquity in Europe of the *Neanderthal*-type long face, big nose, and flattened bulge at the back of the head. The oldest fossil now referred to as *Neanderthal* (Le Biache, France) was discovered in 1976 and is about 19,000 years old, while older fossils (for example, Arago in the Pyrenees) with some *Neanderthal* characteristics, date to the 300,000s or older. Newer, more precise, radiocarbon dates from the end of the *Neanderthal* times, show that, in particular areas, the transition from *Neanderthal* to *Cro-Magnon* was quite abrupt. A *Neanderthal* from St. Cesaire in France, found in 1979, is about 35,000 years old, while the *Cro-Magnon* fossils probably date to at least 34,000, based on comparisons with the Pataud site next door. Such an abrupt transition does not leave enough time for evolution to have occurred in place.

The oldest modern human fossils and archaeological sites of the Aurignacian culture of *Cro-Magnon* are found in Eastern Europe just before 40,000 years ago, while *Neanderthals* still lived in the west, just what one would expect if modern humans invaded Europe from Africa via the Near East. And in the Near East itself, modern humans from Qafzeh, in Israel, excavated in the 1960s, have been dated to ca. 92,000 years ago by thermoluminescence on burned flints.

One problem in the Near East remains the chronological relationship of the Qafzeh modern humans to *Neanderthals*. What might explain *Neanderthal* dominance of this region after a brief period of modern human occupation at 92,000 years? One possible answer lies in the tiny bones of birds, rodents and insectivores found with the human fossils. Tropical African birds, mice, voles, and so on accompanied earlier modern humans, while cold-adapted animals accompanied later *Neanderthals* from Eurasia.

If *Neanderthals* were the cold-adapted archaics, and the earliest modern humans were tropical, this shifting pattern implies that the distribution of the two populations was originally limited by ecological considerations, and that the Near East represented a boundary zone that shifted as the world's climate changed. By 40,000 years ago, when modern humans returned to dominate the region, they seem to have invented a way to get around this ecological limitation. The animals found at the post-40,000 year-old modern human sites remain primarily cold adapted.

It is argued that some interbreeding took place between archaics and early modern immigrants, with modern morphology being dominant and ultimately more successful. This suggests that perhaps some *Neanderthal* genes survive in Western European populations, etc. There is also evidence for modern morphology in some late *Neanderthals* in Eastern Europe, and some *Neanderthal* traits in early modern populations. Yet others point to a number of unique *Neanderthal* traits never found in modern humans that show no evidence for breeding between the two populations.

African Eve Hypothesis

That humans were modern in appearance in the tropics

long before these characteristics appear in Europe seems confirmed by the new dates and data. Of interest is the relationship of the first modern men in Africa to the later ones who occupied Europe after 35,000 years ago.

Cann and co-workers (1987) proposed that a recent migration "Out of Africa" within the last 200,000 years had totally replaced all other human populations. None of the "archaic" East Asians or the *Neanderthals* of Europe had left any descendants at all. All modern humans share a recent African ancestor; this is supported not by the fossil record, or by the dating, but is supported by analysis of the genetic (DNA) differences among the present day living populations. A family tree of the human genetic similarities, based on mitochondrial DNA, reflects only the female ancestry, hence the "Eve" hypothesis.

This last ancestor of all human beings is thought to have been African because Africans are more variable in their DNA than the people of other continents; this suggests that they have been in their native place the longest. Some genetic variants are unique to Africa, whereas all the variants on the other continents are found in Africa too. If *Neanderthals* from Europe or the *Homo erectus* from China contributed to the ancestry of the modern humans, where is their unique DNA?

Family trees based on similarities in the genetic make-up of the Y-chromosome reflect only male ancestry, since women do not have a Y-chromosome. The same pattern was observed, viz. greater variability and unique patterns in the African populations, but no unique patterns outside that continent. The most variable DNA in both the studies belonged to the small isolated populations of the hunter-gatherers in the Kalahari desert (Kung) and Zaire forest basin (Mbuti, Aka, Efe) respectively.

It has been estimated that the human mtDNA diversifies from a common ancestor at a rate of 2-4 per cent per million years. Since the total amount of the difference observed in the modern populations was only about 0.57 per cent, it suggests a time scale of 140 – 290,000 years since all humans last shared a common ancestor.

Recently, the family tree itself has been questioned on statistical grounds; computer programs today can also generate alternative trees in which Africa plays a diminished role. The genetic basis for the total replacement of all the previous human populations by the descendants of "African Eve" appears to be in doubt, although this does not negate the importance of the early fossil evidence from Africa.

Ancient Africans, Whose Ancestors?

Till the early 1970s, *Neanderthals* were generally grouped with other archaics as having modern brains and large primitive faces (and teeth). Western European *Neanderthals*, whose faces were longer and more projecting, and whose elongated heads appeared to have an "occipital bun" of bone at the back, were simply more extreme than others.

Trinkaus (1970) suggests very strongly that all *Neanderthals*, including those from the Near East but not the "archaics" from the tropical environments and East Asia, shared a common and very unusual "post-cranial" form. Their bones, even the fingers and toes were extremely thick and bore heavy markings for the muscle attachments that could not be duplicated in modern samples of skeletons. The joint surfaces were sometimes twice as large as the modern human average. Discovery of a pelvis from Kebara, in Israel, suggested that the way the body was carried was quite different, as the spinal column was deeply indented into the back than in us. However, a hyoid bone from the same site suggests that the movement of the throat, tongue, and voice box in producing speech was similar to ours, despite the greater distance in *Neanderthals* between the neck and the back of the throat.

Neanderthals, in addition, like other cold adapted animals had very large deep chests and short lower arms and legs, to better conserve body heat. Further studies of the face suggested that the very long projecting face and huge, broad nose were distinctive; other large-faced archaics from Africa or East Asia had shorter, flatter faces, with more angulated cheekbones. The differences between *Neanderthals* and other archaics appeared quite striking.

Spanish anthropologists reported the discovery of a unique adolescent male specimen, dated at over 780,000 years; it is the oldest confirmed European hominid. This has characteristics common both to *Neanderthal* and modern man: a bulky mandible, primitive teeth and a ridged brow, all typical of the *Neanderthals*; and human-like cheekbones, depressions on either side of the nose (canine fossa), all typical of modern man. It is not yet clear what species they belong to, although the discoverers have named them as *Homo antecessor* (Bermudez de Castro *et al.,* 1997; Kunzig, 1997).

In Africa, several fossils are intermediate between archaics and moderns. Even the early moderns themselves at Klassies River Mouth are more robust in their limbs than *Cro-Magnons* from Europe. Transitional or even archaic *Homo sapiens* fossils from Asia are quite rare. Regional continuities in Asia, however, are striking to proponents of the multiregional evolution theory. If the earliest modern Asians came from Africa, why do the earliest ones we find already have the flat upper faces and dental characteristics of Asians today? Why are the earlier archaic Asians also flat-faced? "Out of Africa" theorists (Stringer) argue that the flat faces and other features are either primitive features retained in that population, or simply adaptations to the cold dry Asian climate that are favoured each time a new human population reaches the area.

There is evidence that African populations with greatest genetic diversity have accumulated mtDNA mutations for longer time, and are ancestral populations in Asia and Europe, which split off more recently from African roots.

Early Dispersal and *Homo sapiens*

If early *Homo* was able to leave Africa before 2 mya (and later), did they contribute to the prehistoric distribution of *Homo sapiens*, our own species? The new evidence must be discussed in terms of the ongoing debate over "Multiregional" or "Out-of-Africa" origins for *Homo sapiens*. The former argument would use the new evidence to suggest that the early dispersals of early *Homo* set a complex stage for the *Homo sapiens* to emerge at connected points across much of the Old World. However, the new evidence suggests that both the early and later species of *Homo* had the ability to disperse across the continents. Given that *Homo sapiens* fossils appear to be much older in Africa than on any other continent, the answer seems obvious. Having emerged in Africa, *Homo sapiens* dispersed to Eurasia, replacing older populations of *Homo*.

Some believe that there is little question that *Homo sapiens* emerged in Africa, the date of emergence, the technological associations and the dates for its Eurasian dispersals are debatable. It appears that our species originated between 200,000 and 100,000 years ago, somewhere in sub-Saharan Africa. Some recent discoveries in Zaire of ancient and early crafted tools (such as barbed-bone harpoons) indicate that the technology associated with this emergence may have been very advanced indeed, resembling the much later Upper Palaeolithic of Europe. In the Levant, where *Homo sapiens* are evident about 90,000 years ago, a more archaic Middle Palaeolithic technology still held sway. Consequently, we may not yet say whether the European dispersal of *Homo sapiens* was associated with either its emergence or a new technology.

Genetic Evidence for Modern Human Origins

mtDNA comparisons used to trace recent evolution of *Homo sapiens* show that all living humans shared a common ancestor that lived in Africa sometime between 250,000 - 150,000 years ago. The rationale of mtDNA method being that:

- mtDNA is inherited maternally (in cell mitochondria);
- mtDNA is not subject to the same selection pressures as DNA in cell nucleus;
- mtDNA mutation rate is known, and is faster than nuclear DNA, so the differences between populations can be measured within one species that have accumulated recently;
- it provides good information about population migration patterns (demography); and
- African populations with greatest genetic diversity have accumulated mtDNA mutations for longer time and, thus, are ancestral to populations in Asia and Europe, which split off more recently from African roots.

The bulk of the current genetic studies support the recent "Out of Africa" model of modern human origins.

The Story of how we became Human

Despite the protests of creationists and their intellectual allies, science has long taught that human beings are just another kind of animal, but most of the time this seems like a technicality. It's not just the obvious differences – language, civilization, and technology – that set us apart. Even basic biology suggests that humanity has special status. Virtually every other type of animal comes in multiple varieties: dozens of species of monkeys, antelopes, whales and hawks walk, swim or fly the earth, to say nothing of beetles, whose hundreds of thousands of species inspired biologist J.B.S. Haldane's famous quip that God must have had "an inordinate fondness" for them. Even our closest kin, the great apes, fall into four species, divided into several subspecies.

But there's now only one species of humans on the planet, and in the simplified view of evolution most of us have, that's all there has ever been. A few million years ago, most of us think, the half-ape known as *Lucy* appeared in Africa; eventually she begot a less ape-like creature, which evolved in turn into something even more human-like. Finally, after a few more begettings, *Homo sapiens* appeared. Except for that odd side branch known as the *Neanderthals*, the path from proto-apes to modern humans is commonly seen as a succession of new and improved species taking the place of worn-out evolutionary clunkers.

True, we're descended from a creature that split off from the apes millions of years ago. But subsequent events were hardly a steady march from primitivism to perfection. Human evolution more nearly resembled an elimination tournament. At just about any given moment in prehistory, our family tree included several species of hominids – erect, upright-walking primates. All were competitors in an evolutionary struggle from which only one would ultimately emerge. Then came yet another flowering of species that would compete for survival. *Neanderthals* simply represented the most recent version of that contest. And while we'd find it bizarre to share our world with another human species, the fact that we've been alone since the *Neanderthals* vanished some 30,000 years ago is an evolutionary aberration.

The notion that multiple human species are the norm, not the exception, has only got stronger with a series of major scientific discoveries. Since 1994, four new species of hominids have been added to the human family tree, with the latest announced just a few months ago. These date from 800,000 years ago all the way back to 4.4 million years B.P. (before the present).

Scientists have also unearthed new fossils of known

species. This should help them trace the complex relations among our sundry ancestors. One remarkable skeleton, announced this past spring, suggests that modern humans and *Neanderthals* may even have mated successfully. And new evidence of stone-tool use, dating as far back as 2.5 million years, has provided tantalizing clues to how our ancestors thought and behaved.

These discoveries not only further confirm that multiple hominid species are the rule but also bring us much closer to understanding the ultimate mysteries of human evolution: What were the changes that led to modern humans? When did these changes take place, and why? And perhaps most intriguing, will we continue to evolve, or has *Homo sapiens* (wise man) made evolution obsolete?

While all the answers won't be in for some time, experts have identified several key transitions in our evolutionary chronicle. The first, which happened around the time we diverged from the apes, between 6 million and 4 million years ago, was the development of bipedalism – two-legged walking rather than the kind of locomotion Tarzan learned from his adoptive ape family.

The second, which occurred perhaps 2.5 million years B.P., was the invention of tool making – the purposeful crafting of stone implements rather than just picking up handy rocks – and the transition to meat eating. Then, somewhere between 2 million and 1 million years ago, came the dramatic growth of the brain and our ancestors' first emergence from Africa. Finally, just a few tens of thousands of years ago, our own species learned to use that powerful organ for abstract thought, which quickly led to art, music, language and all the other skills that have enthroned humans as the unchallenged rulers of their planet.

Split from the Apes

As recently as five years ago, all that scientists could really tell about our earliest ancestors was when they first appeared. Molecular biologists had measured the differences between human and chimpanzee DNA, then averaged the rate of genetic change over time. By calculating backward, they determined that great apes and hominids branched from a common ancestor between 6 million and 4 million years ago. But no fossils were on hand to support this scenario. The oldest hominid species known, *Australopithecus afarensis* (southern ape of the Afar), could be dated back only 3.6 million years. Its most famous member, Lucy, unearthed in Ethiopia's bleak Afar Triangle in 1974, is a mere 3.2 million years old.

Then, in 1994 and 1995, teams working in Ethiopia and Kenya announced that they had each found a new species of hominid. Both discoveries smashed the 4 million-year barrier. The first – and at 4.4 million years, the oldest – was dug up by an international team in the Middle Awash region of Ethiopia, about 50 miles south of where Lucy was discovered.

All told, the scientists excavated the bones and teeth of 17 individuals. Given their age, no one was surprised that they showed a mix of chimpanzee-like and human traits that as a whole are more primitive than those of *Australopithecus afarensis*: smaller molars, larger canines and thinner tooth enamel, suggesting a diet rich in easy-to-chew fruits and vegetables. The new species, says palaeontologist Tim White of the University of California at Berkeley, a co-leader of the expedition, "is very close to an ape than to an australopithecine and is significantly different from any other hominid."

Because the fossils were too distinctive to be included in Lucy's extended *Australopithecus* family, the researchers called the new species *Ardipithecus ramidus* (*ardi* means ground or floor in the local Afar language, and *ramid* means root). White and his colleagues have since found other *ramidus* fossils at their site but are giving out precious few details until they complete their methodical analysis of the bones. Says *ramidus* co-discoverer Berhane Asfaw of the Rift Valley Research Service in Addis Ababa: "It will be worth the wait."

One prize specimen, they acknowledge, is a partial skeleton found by Berkeley graduate student Yohannes Haile-Selassie (no relation to the Emperor). Alas, the back of the skull is badly crushed. Given the small skulls of *Australopithecus afarensis* and other later australopithecines, however, this specimen undoubtedly had a pint-size brain. At this point in evolution, we are in the minor leagues of brain development.

But the skeleton does include many bones that will help answer the much more important question of how *Ardipithecus* got around. Palaeoanthropologists believe that bipedalism was the first significant modification separating our ancestors from the great apes. By studying the bones and fossil footprints of *Australopithecus afarensis* (Lucy and her line) as well as those of half a dozen other australopithecine species, scientists already knew that our ancestors walked upright long before they acquired other human traits – and that bipedalism gave them a big edge.

According to conventional wisdom, this evolutionary breakthrough came at a time when climate change was transforming eastern and southern Africa from dense forests into open grasslands. Standing upright in such an environment could have offered our ancestors many advantages. It could have let them scan the horizon for predators, exposed less body surface to the scorching equatorial sun (and, conversely, more to the cooling wind) or freed their hands for carrying food or exploring the environment.

But these ideas may not hold good for long. Fossilized seeds, petrified wood and animal bones found by White and his colleagues at the digging site, near the village of Aramis, indicate that it was quite densely wooded when *Australopithecus ramidus* lived there. If the hominid turns out to have been bipedal, as preliminary studies indicate, this could wash away existing theories – though the scientists can't say for sure until other hominid fossil sites of comparable age are found.

Even if *ramidus* didn't walk upright, however, another of the recently discovered human ancestors certainly did. Less than a year after *Australopithecus ramidus* made headlines, Meave Leakey (wife of Richard Leakey) and Alan Walker revealed that they too had found fossils of an ancient human ancestor at two sites near Lake Turkana, in Kenya. Not only is the new hominid very old, dating to 4.2 million years BP, but it is similar in some ways to *Australopithecus afarensis* – though clearly more primitive. Given the family resemblance, Leakey and Walker assigned the fossils to the same genus, *Australopithecus*, and gave the new species the name *anamensis* (*anam* is the Turkana word for lake).

Several of the bones underscore that *Australopithecus anamensis* did indeed walk upright, some 500,000 years before the next oldest two-legged hominid known. But these creatures didn't walk in the modern sense. They were not nearly as efficiently upright as we are, and they had relatively short legs. They had a form of locomotion that we don't know today because there isn't anything equivalent.

Precisely where do *Australopithecus ramidus* and *Australopithecus anamensis* fit into the scheme of human evolution? It is believed that the latter is a direct ancestor of *Australopithecus afarensis* and thus a direct ancestor of modern humans. Some have tentatively labeled the older *ramidus* a "sister species" of all later hominids; it's either our direct ancestor or a close relative of that ancestor. Whatever *ramidus* turns out to be, it's clear that palaeontologists are closing in on the split between apes and humans. "*Ardipithecus* is the closest thing we currently have to the common ancestor of African apes and humans, but its derived characteristics, particularly its teeth, suggest that it postdates that ancestor."

As for the ancestor, hominid fossils have been already discovered that are more than 5 million years old. Leakey and Walker readily acknowledge that they are studying two 5.5 million-year-old hominid teeth and a similarly ancient jaw fragment with an embedded tooth from a site in Northern Kenya. "They look like australopithecines with lots of primitive features," Walker says, but there isn't enough evidence from these fossils alone to claim a new species.

The Earliest Human

Given their 2 million-year-plus life span, the australopithecines were surely one of evolution's better experiments. But nature is an inveterate tinkerer, even with successful species. Between 3 million and 1.9 million years BP, several variations on the *Australopithecus* theme popped up in Eastern and Southern Africa, including *Australopithecus africanus*, *Australopithecus aethiopicus*, *Australopithecus robustus* and *Australopithecus boisei*. (Just to complicate matters, the last three are assigned by some experts to an entirely different genus, *Paranthropus*.)

But figuring out how they arose, how they were related and what they evolved into – those that weren't evolutionary dead ends – has proved elusive. Not only is the fossil record full of holes, but the hominid species from Eastern Africa haven't shown up in Southern Africa, and vice versa. A remarkably preserved skeleton found in South Africa's Sterkfontein cave could change all that. Believed to be at least 3.3 million years old, the bones may belong to *Australopithecus afarensis*, making it the first of Lucy's species uncovered in that area. But the skeleton hasn't been fully excavated yet, and its discoverer, Ron Clarke of the University of the Witwatersrand, in Johannesburg, thinks it may represent yet another previously unknown species.

Whatever the evolutionary relationships between these prehuman species, palaeoanthropologists know that at some point, a second major shift took place. One of Lucy's descendants gave rise to a new kind of creature, the first of the genus *Homo*. Yet none of the known variants of *Australopithecus* seemed anatomically close enough to the *Homo* line to qualify. Some months ago, Asfaw and White's team made another dramatic discovery of a fragmentary skull found near Bouri, an Ethiopian village in the Middle Awash region Northeast of Addis Ababa. Excavated in 1997, its jutting face and upper jaw filled with large teeth clearly belongs to a species more advanced than *Australopithecus afarensis* yet more primitive than the earliest humans. The mix of characteristics wasn't precisely what the experts expected – they were looking to see smaller, more specialized teeth and a larger braincase. So they named their hominid *Australopithecus garhi*. But the skull's intermediate anatomy and its age – about 2.5 million years – put it midway in both time and form between the most recent *Australopithecus afarensis* and the oldest known fossils of our own genus.

That alone would make *Australopithecus garhi* a prime candidate for the long-sought evolutionary link between Lucy's species and the first humans. But the researchers also found that nearby animal bones dating from the same period had been butchered with stone implements. Cut marks on one antelope jawbone suggest that the hominids used a sharp stone flake to remove the animal's tongue. The leg bone of another animal is scarred by cuts, chop marks and signs of hammering, evidence that it was scraped clean of meat and bashed open to expose the nutritious marrow.

Earlier discoveries at Gona, an Ethiopian site about 60 miles north of Bouri, had already shown that someone was using carefully manufactured stone tools in the area at about that time. Now Asfaw and White's team could make a circumstantial case that their species, *Australopithecus garhi*, was the gifted toolmaker. In the 2 million years since the first human ancestor began to walk upright, nothing much had changed. Now something had. Rather than just using sticks and stones to leverage innate abilities – something done by plenty of animals, from chimps to otters to finches – someone had deliberately selected and modified specific raw materials in a sophisticated and consistent way, and with careful intent.

This wasn't just tool use; it was technology. Explains archaeologist Sileshi Semaw, a postdoctoral researcher at Indiana University in Bloomington, who helped find a huge cache of 2.6 million-year-old tools at Gona in the early 1990s: "The Gona hominids [carefully] selected workable raw materials." Since there are no local sources of such materials at Bouri, where the *Australopithecus garhi* fossils were found, the hominids must have carried their tools with them when they travelled there. Did *Australopithecus garhi* make both the tools one used to butcher animals at Bouri? "If it wasn't *garhi,*" then, "what would it have been?" "*Australopithecus garhi* is the best candidate thus far". However, it doesn't rule out the possibility that another species, yet undiscovered, deserves the credit.

Whoever did it, the creation of technology gave its inventors an astonishing advantage over other hominid species. Stone hammers and blades let them exploit carcasses left behind by other predators and permitted them to shift to an energy-rich, high-fat diet. One of the evolutionary consequences of this is the ability to exploit a broader range of habitats, eventually enabling our ancestors to leave Africa and colonize most of the globe. But even more important was the expansion of our brain, with all the potential that went with it; the brain being a very expensive organ in terms of metabolism. It can grow larger only in a species that's routinely consuming high-energy food. One impetus for such growth – and in particular, the growth of the cognitive areas that distinguish ours from other large brains – could have come from our increasingly creative use of tools. Still, the ultimate use to which those big, sophisticated brains would be put would not appear for many hundreds of thousands of years.

Modern Humans

Just as *Australopithecus afarensis* eventually gave rise to the genus *Homo*, so one species came to stand out among the *Homo* line and eventually led to modern humans. The fossil record is far too spotty to say how *Homo habilis* (handy man) and other members of its genus – *Homo rudolfensis*, *Homo ergaster* and *Homo erectus* – were related, to what extent they overlapped or even whether they all represented distinct species. Many scientists believe, though, that it was *Homo erectus* that was the ultimate victor – the direct ancestor of our own species.

Homo erectus was also the first hominid to emigrate from Africa, at least 1.8 million years ago, spreading all the way to China and Indonesia. Then, at some point – for reasons still mysterious - the lineage diverged, with one branch leading to *Neanderthals* and another to modern humans.

An unusually rich trove of fossils has been found at two sites in northern Spain's Atapuerca mountains. One, known as Gran Dolina, has yielded 800,000-year-old hominids that Spanish researchers believe are a new species, *Homo antecessor,* perhaps the most recent common ancestor of modern humans and *Neanderthals*. They have a primitive jaw and prominent brow ridges but a projecting face, sunken cheekbones and tooth development similar to that of modern humans.

Less than half a mile away, *antecessor's* co-discoverer, Juan Luis Arsuaga of the Universidad Complutense de Madrid, is excavating at Sima de los Huesos (Pit of Bones). So far, his team has found, deep inside a natural cave, thousands of fossils from at least 33 hominids of all ages. About 300,000 years old, they appear to represent an early stage of *Neanderthal* evolution. Explains Eric Delson, a professor of anthropology at Lehman College in New York City: "For the first time, we have a good population from a single place and enough variation to show *Neanderthal* features being distilled and standardized."

What occurred some 200,000 years later, when *Homo sapiens* first met their *Neanderthal* cousins – the only other hominid species that hadn't dwindled into extinction – is a matter of much speculation, scientific and otherwise. Our species would end up the only one left standing, but whatever happened to the *Neanderthals* didn't happen quickly. Plentiful archaeological evidence proves that *Homo sapiens* and *Homo neanderthalensis* inhabited the same general turf in many parts of Europe and the Middle East for thousands of years. That doesn't prove, however, that they lived as good neighbours. Populations were so sparse that run-ins probably would have been rare.

A romantic notion of how the *Neanderthals* disappeared has been around for decades: perhaps interbreeding with us eliminated them. May be we all carry a bit of *Neanderthal* in our DNA. Few years ago, molecular biologists tested that hypothesis by extracting some DNA from a *Neanderthal* fossil and comparing it with that of modern humans. Their conclusion: the differences are great enough to rule out significant interbreeding, even though such mating would have been biologically possible.

But a skeleton discovered in Portugal last December gives new life to the old idea. Joao Zilhao and Erik Trinkaus claim that the 24,500 year old remains of a four-year-old child show a mix of human and *Neanderthal* features. The boy could simply be the love child from a single prehistoric one night stand – except that the last true *Neanderthals* had disappeared from the area at least 3,000 years earlier. Ian Tattersall, who says flatly, "It's just a chunky modern kid. There's nothing special about it."

Besides, one isolated case can't explain the demise of an entire population spread across thousands of miles. The mystery is all the greater as palaeoanthropologists learn how similar to our own ancestors, the *Neanderthals* were. They hunted cooperatively, they buried their dead, and their brains were as big as ours. The spec[illegible]' relative equality, says Trinkaus, "makes perfect sense, given that the two groups coexisted for several thousand years without one or the other being dominant."

What may have happened, suggests Tattersall, is that some 50,000 years after modern humans arose, we began using our brains in a fundamentally different way. Despite their burials, for example, the *Neanderthals* left no clear evidence of any ritual or any belief in an after life. Nor is there any hint of *Neanderthal* language. Most telling of all, *Homo sapiens* began, some 40,000 years ago, to create art in an astonishing variety of forms, including cave paintings and female statuettes. All this, Tattersall and others believe, represents a single, profound change: the development of symbolic thought. "Art, symbols, music, notation, language, feelings of mystery, mastery of diverse materials and sheer cleverness: all these attributes, and more, were foreign to the *Neanderthals* and are native to us," he writes in his 1998 book, *Becoming Human.* For the first time, innovation was a routine part of human life that could easily be shared with others – not just something that occurred every million years or so. Against that kind of competition, no other human species could hold out.

The End of Evolution

The development of symbolic thought and complex communication did nothing less than alter human evolution. For one thing, high-tech transportation means that the world, though ethnically diverse, now really consists of a single, huge population. "Everything we know about evolution suggests that to get true innovation, you need small, isolated populations," says Tattersall, "which is now unthinkable."

Not only is a new human species next to impossible, but technology has essentially eliminated natural selection as well. During prehistory, only the fittest individuals and species survived to reproduce. Now strong and weak alike have access to medicine, food and shelter of unprecedented quality and abundance. And technology shows no signs of slowing down, which means that even dramatic changes in the natural world would not necessarily have evolutionary consequences. Wolpoff argues that "we are not going to [adapt to] the next ice age by changing our physical form. We'll set off an atom bomb or set up a space mirror or whatever [to control climate]." Manipulation of the human genome, meanwhile, will eventually let us change the basic characteristics of our species to order. Evolution by natural selection could be replaced, perhaps chillingly, with evolution by human intervention.

That's not to say humanity can't become extinct; a 50 mile wide asteroid crashing down from space would do it. So could a sudden and thorough collapse of earth's ecosystem through pollution, deforestation and the like – unless we establish some colonies in space beforehand. But whatever happens, the long history of multiple hominid species struggling for supremacy on earth is over. After millions of years, evolution by natural selection, operating blindly and randomly, has produced a creature capable of overturning evolution itself. Where we go from here is now up to us.

All in the Family: An up-to-date genealogy of modern humans and their evolutionary predecessors

The more scientists dig, the more hominid species they find. Most are distant cousins that went extinct without progeny; others are our direct ancestors.

Anthropologists are engaged in a debate as to where did modern humans, *Homo sapiens*, originate? Wilson and co-workers (1987) claim that modern humans, a completely new species, arose in Africa alone between 100,000 and 200,000 years ago. This suggests that all modern humans evolved from a single African woman, i.e., descendants of the African Eve spread out from Africa replacing the indigenous human populations all over the world – the "EVE" hypothesis. On the contrary, Wolpoff and Thorne (1991) believe that modern humans evolved from ancient humans gradually in many parts of the world. Another hypothesis, "OUT OF AFRICA HYPOTHESIS" states that the modern humans, *Homo sapiens*, arose relatively recently in Africa and then spread throughout the Old World, replacing the existing populations of earlier forms of *Homo*, specifically the so called "archaic *Homo sapiens*", including the *Neanderthals*. Yet another group of workers believes in the "MULTIREGIONAL HYPOTHESIS", i.e., the modern humans appeared, more or less, simultaneously in Africa, Europe and Asia, evolving *in-situ* from an earlier hominid species, *Homo erectus*, which had migrated from Africa into much of the rest of the Old World about one million years ago.

Before the "Eve", there were two claims about our origins: one that the modern humans evolved at different places at different times, and the second was that they evolved at just one place and then spread out, mixing with the indigenous archaic populations. But the revolutionary aspect of the "EVE" hypothesis was the idea that the modern humans replaced, rather than mixed with the indigenous archaic humans and not that it suggested a single place of origin. The testimony of the fossils suggests otherwise.

Man

Man is one of the primates, a group that originated about

80 mya. This group includes the apes and monkeys as well as the less advanced creatures such as the ancestors of today's tree shrews, lemurs and tarsiers. Primates have large brains, high intelligence, capable, sensitive hands that can grasp well, and stereoscopic vision; the more advanced members of the group have the ability to stand and walk upright. They are social animals, quick to imitate and learn from their fellows, and they communicate well with one another. Primates also have the advantage of being highly adaptable and can move into new areas with different climatic and environmental conditions.

Some 20 or 30 mya, our own ancestors and those of the modern apes – gorillas, chimpanzees, orangutans and gibbons – took to separate evolutionary paths. The apes mainly adapted to a vegetarian life in the tropical forests, while the early hominids, as man's ancestors are called, may have lived in more open country and eaten meat as well as plant foods.

There are three known groups of fossil men. The oldest of these *Homo erectus*, have been found in Java and North China. In Africa, there have been several discoveries of comparable type and antiquity. The second group, the *Neanderthalers*, was widely spread during the late Ice Age, about 30,000 years ago. Contemporary with them, were the known ancestors of modern man, such as the *Swanscombe*. Differentiation of the present races had begun by the end of the Palaeolithic times. Man did not reach America until about 15,000 years ago.

Early *Homo erectus*, the first hominid, is characterized by a lateral nasal aperture eversion – a marked departure from the ape-like nasal morphology, i.e., a projecting nasal skeleton which closely resembles that in modern *Homo sapiens* (Franciscus and Trinkaus, 1988). *Erectus* was the first to leave Africa and settled in vast expanses of Europe and Asia. Almost all the palaeoanthropologists agree that this first great human migration began around a million years ago. *Homo erectus* (400,000 B.P.), an upright and bipedal species of the genus *Homo*, was widely distributed from China and Java to North and East Africa and Europe (?).

Two adult human crania, damaged but relatively complete, discovered *in-situ* from the middle Pleistocene terrace deposits of the Han River in Yun County (Yunxian) in Hubei Province in

China show features resembling both *Homo erectus* and the archaic *Homo sapiens*: flattened face, orthognathus but with moderate alveolar prognathism, maxilla with a canine fossa (Tianyuan and Etler, 1992). These hominid fossils retain many typical features associated with *Homo erectus*, for instance long, low and sharply angulated cranial vault, greatest cranial breadth close to the base. It can perhaps be argued that these Yunxian hominids were highly polytypic and regionally differentiated. Or else, they form an essential part of one evolving lineage in direct ancestry to modern humans.

Thus, the question arises: "Is the theory of an 'African Eve' still valid"? This is difficult to answer as fossils of the pre-modern human *Homo erectus* have been found in most of the same places from where fossils of early modern humans have been reported. Recently, some fossil skulls from Java have been dated to be between 51,000 and 27,000 years suggesting thereby that they are far more recent than previously thought. It means that *Homo erectus* and *sapiens* co-existed in this region for sometime (Swisher *et al.,* 1996). This implies that *Homo erectus* might have migrated out of Africa where it first appeared and then evolved into modern *Homo sapiens* in different regions around the World. Modern *Homo sapiens* could also have first appeared in Africa before migrating along the same routes as of *Homo erectus*. If the data on *Homo erectus* from different places were similar to that of modern humans from the same places, then this would support the contention that modern humans evolved separately in different parts of the World. On the other hand, if the data on *Homo erectus* was very different from all the data on modern humans from the same places, then this would support the idea that modern humans evolved only once; but this would not tell when or where this change occurred.

The late Pleistocene interface between the Neanderthalers and the early modern humans during 40,000 to 30,000 mya greatly facilitates understanding of the evolution of humans in Europe. For instance, the Atapuerca (Spain) fossil remains resemble the 'archaic *Homo sapiens'* but are well-differentiated from the Asian *Homo erectus* group (Stringer, 1993; Arsuaga *et al.,* 1993). By about 3 mya, some of these early hominids had learnt to make and use simple stone tools. During the next 2.5 my, the forms as *Homo habilis* and *Homo erectus* developed increasingly skilful hands which enabled them to use more complicated tools. They walked and stood erect and as they learnt to hunt big game, they increased their intelligence. Towards the end of this period, they had learnt to use fire and they probably developed speech and language. During the last 100,000 years, the picture becomes clearer. First the *Neanderthal* man, a subspecies of *Homo sapiens* appeared about 70,000 years ago and spread widely through Europe and Asia, only to die out some 30,000 years ago, and then appeared the *Homo sapiens sapiens* – modern man. As more human beings colonized different regions of the World and became physically adapted to the fresh climatic conditions, distinctive races gradually evolved until the racial differences, that exist today, became established. There have been no major changes in man's physical form since the end of the last Ice Age 10,000 years ago. During this period, man began to cultivate the land, to domesticate the animals and to live a communal life; the indigenous races of man, thus, evolved.

A partial mandible and an upper incisor from the Longgupo cave site in China are reported to represent an early dispersal of *Homo* from Africa to Asia (Huang *et al.,* 1995). This early species of *Homo* resembles *Homo ergaster* (Wood and Turner, 1995) or *Homo habilis* (Ciochon, 1995) rather than the supposedly first Asian colonizer *Homo erectus*. These Longgupo remains are used to buttress the argument that *Homo erectus* evolved in Asia and is not ancestral to modern humans (Culotta, 1995). However, doubts have been expressed about the human attributes of the mandibular fragment and the specific hominid species affinity of the upper lateral incisor (Schwartz and Tattersall, 1996). Therefore, it is premature to base conclusions on these Longgupo remains.

Palaeoanthropologists are now proposing that all fossil specimens assigned to *Homo erectus* and archaic *Homo sapiens* (including the *Neanderthal*) should be reclassified into a single species, *Homo sapiens*; the species *erectus* being insufficiently distinct from the species *sapiens*. As a result, *Homo sapiens* would include hominids with brains as small as 850 cc (including those formerly known as *Homo erectus*) and modern men with brains having cranial capacity up to 2000 cc (Shipman, 1993; Wolpoff and Thorne, 1991). This contention (argument) is based on a study of the three species of the genus *Homo*: *habilis*, the earliest species; *erectus*, the second in line; and *sapiens*, the modern species. The earliest species of *habilis* was found to be morphologically distinct from the other two, while no anatomical markers were found to be consistently separating *erectus* from *sapiens*. Two newly discovered fossils from Yunxian in China show mixed anatomical characteristics: construction of the braincase and the base of the skull resemble that in *Homo erectus* but their face and 'brain' sizes are closer to that of the *Homo sapiens*. These skulls are so badly crushed that interpreting them, however, is fraught with difficulty and dogma: firstly, comparing Asian *Homo erectus* with Caucasoid *Homo sapiens* is inappropriate and misleading. Secondly, *Homo erectus* skulls from different times and places vary enormously; for instance, African *Homo erectus* is 1.6 my old whereas the Indonesian *Homo erectus* is 700,000 years old. The former has no sagittal crest/keel but has a raised forehead whereas the latter has a sagittal crest and a receding forehead.

Assuming that the divergence between *Neanderthal* and modern man was probably only about 200,000 years, our differences must have been remarkably slight. Simple calculations indicate our genetic constitutions must have differed by less than 0.1 per cent. Qafzeh and Kebara fossils not only support these contentions but do show that both overlapped for at least 40,000 years in the Middle East. Both, the *Neanderthal* and the modern man spent their lives hunting, gathering fruits and berries, and making stone tools. About 45,000 years ago, the contemporary humans produced innovative, fully modern people who proceeded to spread westwards into Europe quickly supplanting *Neanderthals* (McKie, 1992).

One remarkable 24,500 years old skeleton of a young boy found in Portugal suggests those modern humans and the *Neanderthals* may even have mated successfully (Duarte *et al.*, 1999). And new evidence of stone tool use, dating as far back as 2.5 mya, has provided tantalizing clues about the 'behaviour and thinking processes' of our ancestors.

However, the mtDNA of the *Neanderthal* differs by a surprising amount from the equivalent modern human DNA suggesting that the *Neanderthals* were not particularly closely related to any modern humans. This finding implies that they were a different species (Krings, *et al.*, 1997; Kahn and Gibbons, 1997). Dobson (1998) suggests that the *Neanderthal* features are caused by an iodine deficiency, or by a genetic difference in the thyroid, for instance, goitre and/or cretinism.

Recent genetic data, both nuclear and mtDNA, suggest that modern humans originated in Africa, within the last 200,000 years. However, the fossil records of Asia and Australia still have many surprises. Stringer (1990) believes that even if our true ancestors were still restricted to Africa 130,000 years ago, Asia must have been the primary dispersal center for subsequent radiation of human beings around rest of the World. There is a growing belief that the last common ancestors of humans, chimpanzees, and gorillas was a robust-jawed ape with thick dental enamel, which inhabited Africa from 10 to 5 mya (Wolpoff, 1983). Thus, the discrepancy between palaeontological opinion and the molecular findings has narrowed considerably.

Man probably appeared on earth in between the last two great continental glaciers, for instance, from 1,50,000 years to 50,000 years ago (Howell, 1960) and began to differentiate into races at the same time.

The sapients: Neanderthalers – inflated face, retromolar space (gap behind the last molar tooth), and the occipital bun.

The *erectus/sapiens* transitionals: Skull from Omo region of Ethiopia (Omo II) is very fascinating – cranial capacity is virtually up to modern standards, slope of forehead and nuchal plane is very archaic. Is it erectus or modern?

Another skull (Omo I), of the same age, differs in many ways. Are the two skulls belonging to the same species? They are distinguishable at least in grade, if not in clade (Durant, 1989).

The Pre-*Neanderthalers*: Morphological details of the hipbones of the *Arago* skull dated at 1.9 mya (ER 3228) and 700,000 BP (OH 28) are similar – similar features persisted for as long as over million years – morphological punctuation event.

The latest genetic and linguistic research, however, suggests that today's various racial groups might have descended from people who colonized the World just over 100,000 years ago, instead of (rather than) evolving separately perhaps a mya.

About 35,000 years ago (Late Pleistocene), modern man, *Homo sapiens sapiens*, emerged in Europe; his ancestors may

About 35,000 years ago (Late Pleistocene), modern man, *Homo sapiens sapiens*, emerged in Europe; his ancestors may have been the early *Neanderthal* or even earlier forms of Man, such as *Swanscombe*. Modern man had a more varied set of bones and stone tools and he created the World's first known art, the engravings and paintings, on the walls of the cave shelters in Southwest France and Spain, and the Venus figurines found on sites, such as Willendorf in Eastern Europe.

During the last 30,000 years, Man had spread across the globe reaching North America from Asia across the land bridge that joined the two continents where the Bering Strait now flows. The Mongoloid people then crossed the Bering Strait in a series of migrations; the most recent of which was that of the Eskimos. Man may also have reached Australia at about this time when the continent was still joined to the Asian mainland.

Diamond in his latest book, *The Last Chimpanzee*, argues that modern humans became fully modern in their behaviour rather suddenly about 40,000 years ago. This 'great leap forward' or 'human revolution' is largely based on the perspective from Europe, where major changes in technology (blade and bone tools); economic strategies (ambush hunting, fishing); size of social networks; and symbolic activities (art) occurred over a few hundred years as the *Cro-Magnons* replaced the *Neanderthals*.

The discovery of new sites, fossils and data dating to between 250,000 and 40,000 ya have accelerated since the 1960s. Even with the limited exploration of Africa to date, it seems that, like modern human facial shape, some of the modern behaviours of the 'human revolution' appear well before 40,000 years ago in Africa. For example, the later *Neanderthals* chased their prey and stabbed it with sharpened sticks or an occasional stone-tipped spear. On the other hand, the Central and Eastern Africans hafted small delicate stone points on to the spear (or even arrow-shafts, made stone blades, backed triangles or crescents, barbed stone points, and other bone tools) and engaged in regular fishing and ambush hunting. They ground their food (and some pigments) with grindstones; scratched designs on ostrich eggshell fragments; and traded precious raw materials such as obsidian over more than 800 kms. Like later *Neanderthals*, the early modern humans also buried their dead with grave goods (Brooks, 1992). The early anatomically modern populations in Africa and the Near East used a technology very similar to that used by the *Neanderthals* in Western Europe, viz. a Middle Palaeolithic technology, using Levallois flaking techniques and producing a simple tool kit made on flakes (e.g., Mousterian culture).

Genetic differences indicate that the *Neanderthals* were a different species than the early humans who swept them aside in Europe and Western Asia – although they appear to have split from a common ancestor a half-million years ago. Comparing 378 base pairs of the *Neanderthals'* mtDNA to that of modern humans, an average of 27 differences have been found between modern and *Neanderthal* DNA – far more than the typical variation of eight among modern humans. It has been conclusively shown that the *Neanderthal* DNA sequence falls outside the normal variation of modern humans (Paabo, 1997). Even if the *Neanderthals* were not our ancestors, they were strikingly similar: they walked erect, used tools and there is evidence that they co-existed and learned some skills from *Cro-Magnon* people. However, there is one striking difference, namely that they were bigger than modern humans and had larger brains.

By 50, 40,000 years ago, new data show that Africans wore beads of ostrich eggshells, and engaged in organized mining of precious raw materials. Elsewhere, modern humans had used boats to reach Australia, New Guinea, and New Caledonia, where rock art has been dated to 32,000 years ago. Outside of Europe, the 'great leap forward' began earlier and was more like a slow jog, with occasional detours and backward movements.

But were the *Cro-Magnons* African?

Although modern humans appear to have developed earlier in Africa, evidences do not demonstrate migration of modern humans to Europe. Despite earlier claims for the fossils from Grimaldi, Italy, African characteristics such as nose shape and width, wide distance between the eyes, and forward projection of the mouth, do not occur in the early Europeans. Grimaldi itself is not only "African" but is considerably later in time than the earliest modern Europeans – new dates suggest an age of archaeological sites, the Aurignacian culture of the *Cro-Magnons* appears first in Central and South-eastern Europe, just before 40,000 BP, spreading to near Barcelona, Spain by ca. 38,000 and finally to France and Germany by 34,000. Southern Spain, near the straits of Gibraltar, is one of the last areas to make transition from Mousterian thick scrapers; and bone points of the Aurignacian are quite unlike anything from the preceding Mousterian culture of *Neanderthals*. So, it was assumed that it came into Europe from outside. Yet there is nothing 'outside' in this time range, either in the Near East or in North Africa, from which Aurignacian can be derived. In much of Africa and the Near East, at ca. 40,000, the stone industries were characterized by finely-made small blades, many with narrow points created by blunting or battering the sides, or by small points with a tang or projection for hafting. The Aurignacian does show up in the Near East, but recent dates suggest that this is only after it is well-established in Europe, at about 34,000 ya. The Near East may have been a migration corridor, but it was open in both directions.

Cultural Evidence for Modern Human Origin

Even the early anatomically modern populations in Africa and the Near East used a technology very similar to that used by *Neanderthals* in Western Europe: a Middle Palaeolithic technology, using Levallois flaking techniques and producing a simple tool kit made on flakes, viz. Mousterian culture. For instance, Klassies River Mouth site in Southern Africa, with evidence for simple flake tool kit, opportunistic hunting and shell-fish collecting, use of hearths, associated with fossils of early anatomically modern humans at 120,000 – 100,000 BP.

The first signs of modernity that are easy to spot are associated with Upper Palaeolithic Cultures (40,000 – 15,000 BP), viz.

- appear across Old World by 40,000 BP (perhaps older in Africa);
- associated with technological innovation (new raw materials used, like bone, antler, etc., new stone flaking techniques, such as core blades, soft hammer percussion, heat treating flint), and more deliberate control/design of artefacts;
- greater diversity of specialized tool functions (e.g., needles, harpoons, engraving tools, large bifacial points);

- stylistic variation in material culture: distinctive styles develop in different regions, and change more frequently through time (e.g., Venus figurines across Europe);
- ornaments and art appear for the first time in many parts of the world, including Europe, Africa, and Australia; and
- such art seems to have symbolic/ritual significance that we can only speculate.

Cultural variation (for instance, in art styles) suggests that fully modern language had developed by the Upper Palaeolithic times, as people were using material culture to communicate with each other (e.g., symbols of identity, ethnicity?). Upper Palaeolithic folks in different parts of the world became ecologically sophisticated and very successful. For example, at hunting migratory herd of animals and fishing for seasonal salmon at sites in France, to the point where Upper Palaeolithic hunting success may have helped drive many large animals common in Ice Age Europe (and Australia, Africa and North America) to extinction.

Social co-operation is evident in such hunting tactics, such as the collaborative effort needed to build mammoth bone houses in Ukraine (e.g., site of Mezhirich). Some have even suggested that key cave art sites like Lascaux in France, or Alta Mira in Spain, would have been the ritual sites associated with social co-operation; a way of coping with survival in Europe during the Ice Age. People invaded new lands with new skills: for instance, Australia by 50,000 years ago using boats; Siberia by 30,000 years ago; North America by 18,000 years ago via Bering Land Bridge.

Rethinking?

A new species of early human, *Australopithecus bahrelghazalia* (the species name is based on the Arab name of a nearby riverbed) has been reported from the central African desert area of Chad (Brunet, 1996). This site is so far away from the traditional ancient fossil sites that it may change the understanding of where human ancestors originated; the remains discovered in the Chad desert were found 2.400 kms to the west of contemporary fossils from the same lineage found in Ethiopia. The remains found by the French team of palaeoanthropologists comprise of a lower jaw with seven teeth and a separate canine tooth of a hominid that they believe lived 3 m to 3.5 mya. The specimen from Chad is different enough from the known hominids to plausibly call it a new species (Pilbeam, 1996) and it opens up a major new area of research in Africa. The distance between places in east Africa and South Africa where the oldest fossils were found is similar to the distance between Ethiopia and Chad in Central Africa. The find in Chad adds a new dimension about the place of origin of human ancestors and how far they ranged. This discovery if challenging conventional thinking but it certainly suggests that the cradle of humanity is much larger than thought earlier.

While the discoveries of the past two decades have gone far towards demonstrating the priority of continents other than Europe in the evolution of modern humans, the data also suggest that this was not a simple event of evolution followed by migration in one direction. Replacement of earlier populations may not have been total. More and better dates and data, particularly from regions such as western Asia, Turkey and the Balkans, as well as Africa, may go far towards clarifying the complex interactions involved in this transition.

The discovery of a common ancestor, of both the *Neanderthals* and modern Man, *Homo antecessor*, suggests that the human family tree had many branches (rather than being linear). The present day humans shared ancestors with other human-like beings who eventually became extinct. This hominoid first evolved in Africa, where it gave rise to modern man, and thereafter migrated to Europe about 1 mya.

The new find of Eosimia fossils (Gebo, 2000) from a limestone quarry at Shanghuang (about 100 miles West of Shanghai, China), adds to the growing evidences of a rich evolutionary history of primates in Asia. This raises questions about the long-held view that human origins probably can be traced to Africa. The oldest remains of human ancestors ever found are between 2 to 4 my old, and are found in East Asia.

In linking the early dispersal of 'early' *Homo* with its emergence, we are describing a hominid very different from the australopithecines whose bipedal gait but still the ape-like anatomy must have limited them to forested locales. Thus, the significance of an early dispersal to Asia is manifold. First, the climatic conditions of cool aridity that played a great role in the emergence of *Homo* itself also drew hominid populations out of Africa and into Asia. The emergence and dispersal is, to a great extent, a product of the environmental change. Nevertheless, early *Homo* emerged with a radical, yet still generalized, set of characteristics that granted it ecological hegemony across the subtropical Old World. An early intercontinental distribution signifies a hominid not adapted to the specific territorial conditions, but adapted to manage many local conditions through physical presence, technology and flexible social organization. Ironically, as the first species to use technology, early *Homo* colonized much of the subtropical Old World without the benefit of language, symbolic culture or individual consciousness, as we know it.

10

Molecular Clock

The nature and timing of the evolutionary events that have led to the extant primates, especially to humans and the African apes, can only be partially clarified by the molecular clock hypothesis: the amount of genetic change being related to the dates of divergence. As, a majority of the silent mutations are not revealed in the amino acid sequence data, their detection should eventually produce sufficient data for a more thorough test of the clock concept. If silent mutations constitute a large proportion of change in a molecule, and if generation time is not proven to be a significant factor affecting rates of molecular evolution, then clock dates based on nucleotide sequence data may fall more closely in line with divergence dates suggested by the fossil record. Generally, an accurate clock would require that neutral mutations and random walk have been the most important processes responsible for the evolution at the molecular level. Functional analysis of α– and β–hemoglobin and cytochrome C have demonstrated that selection has been crucial in orchestrating change in the history of the molecules.

A striking example of an accelerated rate of amino acid replacement comes to light in the primate region of the maximum parsimony tree of cytochrome C sequences. After the acceleration in the early primate stem to the catarrhine ancestor [10 to 20 nucleotide replacements (NR) per cent], a marked deceleration of rates occurred. The higher primate lineages evolved at a rate of only 1.5 to 2 NR per cent.

If we regard *Australopithecus* as more closely related to *Homo* than to *Pan*, one finds that during the last 5 my, the rate of globin evolution averaged over the sequences of six non-allelic loci – and, thereby, representing a total of 30 my of descent – is only 2 NR per cent. The rate of globin evolution in the terminal stage dropping to less than 1 NR per cent. The sequence data on amino acid residues – from six globin chains, carbonic anhydrase, cytochrome C and fibrinopeptides A and B – shows that chimpanzee and man differ by only three amino acids. The maximum parsimony evidence from fibrinopeptides A and B, and myoglobin sequences supports the placement of the chimpanzee and the gorilla genealogically closer to man than to the orangutan or the gibbon. But evidence from myoglobin sequences alone places this hominine assemblage of African apes and man closer to the orangutan than to the gibbon, and evidence from the myoglobin sequences alone places the hominines closer to the gibbon than to the orangutan. Parsimony evidence from carbonic anhydrase sequences shows *Pan* and *Homo* to be closer to each other than to the *Pongo*.

Evidence from fossils strongly indicates that the ancestral splitting between *Homo* and *Pan*, the terminal stage of human descent lasted at least 5 my, and possibly as many as 15 my. The profound showdown in rates of protein sequence evolution during this period resulted because a higher density of co-adapted molecular specificities related to more prolonged and complex ontogenes had evolved in man's ancestors. Not only were these ancestors subjected to more intense stabilizing selection, but also their internal environments were better buffered against assaults from the external environment. Thus, they were more immune from the positive selective pressure from new mutations (Goodman, 1977).

Chromosomal Evolution

The very instant life appeared on earth in the form of blue algae, evolution began. Evolution, whether necessary or not, has produced in three billion years, the numerous species which are now extinct. It is well-established that during the course of evolution, differentiation of species into new ones was accompanied by changes in the gene pool and often in the karyotype, which finally leads to the genetic isolation of species and the failure of getting offsprings from them. Structural chromosomal rearrangements occurring in the course of species evolution can now be fully determined by the innovation of the recent specific cytological staining techniques, f.i. banding techniques. Even polymorphisms involving apparent additions of one or more segments to the chromosomes are well-demonstrated by the longitudinal differentiation of each of the chromosomes (Seth and Saxena, 1977). Since chromosome features are derived characters, it becomes axiomatic that chromosomes may be used to reconstruct phylogenies, because "a phylogeny does not need to be based on fossils but can be inferred upon a careful comparative analysis of morphological characters" (Mayr, 1974).

Chromosomal rearrangements, that are believed to play an important role in evolution, are fusions of the acrocentric chromosomes or Robertsonian translocations. They reduce the number of chromosomes, and many phyla are known in which one species derives from another by reduction in their chromosome number; other types of rearrangements have been postulated but not proven without dispute. The mechanisms that seem to have predominated in the better known evolutionary series, at least in the vertebrates, are Robertsonian translocations (fissions and fusions), and inversions. The recent banding techniques have further substantiated the role played by centric fusions. For instance, one such fusion has occurred in the human phylum reducing the chromosomal number from 48 in apes to 46 in man.

The mechanisms of chromosome repatterning are an inferred class of phenomena whose consequences may be observed under the light microscope, and thus are distinct from point mutations of nucleotide triplet sequences, as these latter occur at sub microscopic level. The behaviour and consequences of these mechanisms is manifest in their meiotic products that are lethal or sublethal. These are of interest to others because of the consequences of fetal wastage, birth spacing, and differential fecundity. Systematists, however, are concerned only with the viable chromosome repatterning event and its evolutionary consequences. It is this event that seems to contribute significantly to the reproductive isolation of demes and species over time.

The chromosome rearrangements retained by evolution appeared in single individuals or in couples; these rearrangements had a major function, namely to create reproductive barriers capable of isolating new species from old species. An inversion modifies the existing order of genes within a chromosome; this has greater implications on the "position effect". The linear order of genes with regard to the centromere is important for their translation. Inversions, which change this order, would than have great modifying values, for the gene may have different activity if it changes its neighbours. The resulting effect on speciation would be paramount.

Closely related primate species are separated from each other by chromosome rearrangements that constitute the necessary reproductive barriers. After these barriers were operative, species evolved on their own by acquisition of gene mutations that are responsible for the biological personality of each species.

If one considers closely related species, then they should produce the same proteins, or practically the same. It is well-established that Man and the chimpanzees manufacture protein molecules that are more than 95 per cent identical. In other words, man and chimpanzee have practically the same chromosomes, they carry the same genes on the same chromosomes, and they manufacture practically the same proteins. Yet they are different in many respects. This variation can be attributed to the different organizations, in higher organisms, of the genetic material within the chromosomes, the existence and the role of redundant DNA, the importance of genetic regulatory mechanisms, the role of heterochromatin, and the isochromosome communication network, etc. It is not the nature of the molecules produced by the species, but the way they are produced and utilized in the metabolic pathways that differentiate species from one another.

Phylogenetic trees of chromosomal evolution deduced from the study of Man and the anthropoid primates, are strongly reminiscent of other chromosomal evolutionary patterns, such as those observed during carcinogenesis, and known as *clonal evolution*.

Chromosomal Homology

Unlike most genetic comparisons, variation in the chromosomes can be seen directly through the microscope. Chromosomal comparisons also differ from other genetic analyses in that they are generally qualitative and not quantitative: certain important features are present or absent throughout an entire species. Chromosomal studies contribute to a better understanding of the phylogenetic relationship between different taxonomic groups than does their morphology. Comparative cytogenetics of species, at the generic level, provides an additional means of assessing the phylogenetic relationship and also in establishing the mechanism of chromosomal evolution. Rearrangements, both at the chromosomal and the DNA levels, have been an important factor in the origin of the species and their polymorphism along with the associated phenomenon like recombination, selection and isolation. The structural changes in the chromosomes that might have occurred during the evolution of the species can now be easily identified by the different banding techniques resulting in the longitudinal differentiation of the chromosomes.

Comparison of the chromosomal banding patterns of different species has revealed that pericentric inversion, besides Robertsonian translocation, plays an equally important role in generating reproductive isolation (Egozcue, 1969). This is particularly true in the Platyrrhini, where the results of independent action of Robertsonian translocation or fusion and pericentric inversion of the same ancestral karyotype lead to the formation of two species. Ciochon and Chiarelli (1980) have suggested the possibility of an African origin for the Platyrrhines.

The pattern of cytogenetic evolution in the Catarrhini (the group indigenous to the Old World) and the mechanisms leading to their domain of the extant progeny(s) (plotted array of related karyotypes) are unfortunately not as dramatic as in the rodent lineages, and the least obvious of any of the primate suborders.

Chromosomal repatterning within the Catarrhini shows that the most dominant slope relates to 2n = 48 and at least ten genera fall "on or near" this slope: *Pongo, Pan*, *Gorilla, Homo, Presbytis, Nasalis, Allenopithecus, Symphalangus* and *Hylobates klossi* (the latter two both have 2n = 50). A Matthey's Line of Fn=98 characterizes the second 'slope' and at least eight genera fall "on or near" this slope: *Nasalis, Allenopithecus, Symphalangus, Hylobates* (*klossi, concolor*), *Erythrocebus, Miopithecus*, and *Cercocebus aethiops*. The third possible slope is described by a horizontal cluster of populations having a "constant" number of 0 to 2 telocentrics: twelve genera fall "on or near" this slope: *Symphalangus, Hylobates, Allenopithecus, Nasalis, Presbytis, Rhinopithecus* and six genera comprising of 42 chromosomes (2n) – *Macaca, Papio, Theropithecus, Mandrillus, Cercocebus* and *Cynopithecus*. A fourth slope of 120° represents the domain of extant progeny of many species of *Cercopithecoidea* (Mai, 1983).

Gibbon chromosomes have been considered generally non-homologous when compared directly to other hominoid taxa (de Grouchy *et al.,* 1978). Pericentric inversion has been the major repatterning mechanism involved in the gibbon evolution. Mai (1983) suggests that the karyotypes of the lesser apes reflect a dramatic deviation from the more conservative evolution seen in the great apes. Their gross chromosome morphology is convergent with certain cercopithecoid taxa. Because of the large differences seen between the lesser and great ape karyotypes, however, it is not presently possible to suggest a point of reticulation of the lesser apes with a cladogram of the great apes. Excluding the gibbons, than, it is of interest to infer the cladogram for the rest of the Hominoidea at a higher level of resolution.

The karyotypes of the great apes and other non-human primates are of special interest because of the evidence they provide concerning chromosomal changes during the evolution of the human species. The Pongidae are the most studied family, and among them, the chimpanzees have been examined the most. This species possesses chromosomes that are not very different from those of man. Normally, as one descends the evolutionary scale, the homologies based on the banding pattern become less certain and the problem intensified (Miller *et al.,* 1974). But in the case of macaques, one finds homology of almost the same extent with that of man as exist between Man and the Chimpanzee (Seth *et al.,* 1976). This suggests that the chromosomal changes involved in the process of karyotypic evolution in Man, chimpanzee and the macaques are primarily because of the rearrangements of the genetic material within the genome. Although, by and large, it has also been demonstrated that the differences between these three groups are primarily through Robertsonian translocation mechanisms, peri- and para-centric inversions. Most remarkable, therefore, is that the genetic material of their progenitor has been conserved in Man, chimpanzee and macaques, despite their occupying different status within the order Primate. It should be mentioned that the similarity of chromosome banding patterns between species does not necessarily imply the presence of genetic homology, for different kinds of banding techniques often result in different staining reactions, especially to constitutive heterochromatin.

One of the significant features of the human karyotype includes a Robertsonian translocation of two chromosome pairs thereby reducing the diploid number to 2n = 46 and forming human chromosome 2. The other significant feature is the emergence of C-banded regions below the centromeres of chromosomes 1, 9 and 16 and on the q- arm of the Y-chromosome. The common ancestor of chimpanzees and gorillas passed on the unique features to both its daughter species. Most striking is the set of terminal C-bands in the karyotypes of chimpanzees and gorillas. Interestingly, this character has evolved independently in a common ancestor of the two gibbon species, viz. *Hylobates concolor* and *Hylobates syndactylus.* Probably, this feature links these two species as one another's closest relatives. Yet another feature, i.e., inversion of chromosome 12, also distinguishes chimpanzees and gorillas from humans and orangutans. In addition, several inversions are unique to the karyotype of each species, but only one translocation is yet established in this group, namely, an exchange between chromosomes 5 and 17 in the gorilla.

Cladistic representation of the dynamic hominoid population changes network concludes that during the course of hominoid evolution, population's homokaryotic for newly arisen morphs budded from parental populations (Wagner, 1969). Several of these chromosomal morphs have subsequently become extinct, i.e., do not appear as extant populations without additional changes, and have been replaced by chromatids with the repatterned linkage groups observed in existing hominoids. It is, however, useful to use the human homologue as a reference when comparing the chromosomes of apes and Man. The chromosomes of chimpanzee, gorilla and orangutan (which appear to be homologous, for instance, to human chromosome 14) are also designated as 14 even though the specific chromosome involved is 15 of the chimpanzee and orangutan and chromosome 18 of the gorilla.

The chromosomes of the gibbons are very highly derived and as such it is impossible to assign them in a phylogenetic reconstruction of the superfamily Hominoidea based only on chromosomes. It would, however, be more appropriate to compare the karyotypes of the great apes with those of baboons and macaques; their karyotypes have not changed greatly since the OWMs diverged from the hominoids. However, in these karyotypes many of the human chromosomes can be found almost intact.

Baboons and macaques, and most of the gibbons as well, have only one pair of the metacentric chromosomes bearing NORs. All the great apes, i.e., the chimpanzees, gorilla and orangutans, have several pairs of the acrocentric chromosomes with NORs. Thus, it appears that the radiation of the great apes during the Miocene epoch was accompanied by a 'colonisation' of the various chromosomes by nucleolar organizers. This was accompanied by a fission of three pairs of chromosomes that raised the diploid number 2n=42 to 2n=48.

The branching of the orangutan from the lineage leading to humans, chimpanzees and gorillas was associated with the emergence of the chromosomal regions that stain brightly on being treated with quinacrine. These differ from the ordinary Q-bands. The "Q-brilliant" zones are restricted to the chromosomes of Man, chimpanzees and gorillas. They probably reflect localizations of repetitive DNA. Significantly, the Q-brilliant material differs in its chromosomal localization in each species. Man, chimpanzees and gorillas also have fewer NOR-bearing chromosomes than does the orangutan. A common African ancestor must therefore have lost the NORs from chromosomes 2 and 9.

Most of the primate species have now been karyotyped. Of these, the great apes have been studied in great detail, given its own karyotype and system of numbering.

Miller's (1977) hypothesis shares some morphological characters on some homologous chromosomes that the gorilla is more closely related to humans than is the chimpanzee has been falsified since her hypothesis is incompatible with the branching pattern produced by the cladistically more robust chromosomal repatterning data (Mai, 1983). More than 30 structural rearrangements have become fixed in the separate lineages leading to the living hominoid species. Of these events, at least 23 are inversions.

DNA

All the chromosomes of hominoid primates (apes and humans) have small C-bands at the centromere. Additional prominent C-banded regions in Man are on chromosomes 1, 9, 16 and Y. Chimpanzees lack such C-bands, but have others: in the middle of the long arm of the two other chromosomes, as well as at the tips of several chromosomes. Gorillas have terminal bands like those of the chimpanzees, and C-bands on the homologues of chromosome nos. 16 and Y, as in humans. Orangutans have C-bands only at the centromere. The comparative study of DNA replication reveals no differences between the karyotypes of man and chimpanzee. Dutrillaux and his co-workers (1978b) extended this comparison to some Cercopithecidae and found no major changes. It suggests that no changes have occurred in the replication time of chromosomal bands that have undergone a change in their location from one species to another after a structural rearrangement.

With the BrdU incorporation technique, it was found possible to show that all the chromosomal bands do not replicate their DNA at the same time, but in a certain order. Dutrillaux and his group (1976) were able to study the sequence of replication of 272 Q- and R-bands in the human karyotype. This new criterion of chromosome analysis was then applied to the chimpanzee and the bands when compared with those of man showed no differences in their replication (Viegas-Pequignot *et al.,* 1978). No differences were observed in the *Papio papio* except for the juxta-centromeric region of orangutan chromosome 17, which is late replicating. Dutrillaux (1979) concludes that this segment is heterochromatic in nature, which explains why the equivalent of human 17 possesses slightly more material in *Papio* and in *Pongo pygmaeus*. Same sequence of replication of the R- and Q-bands was found in other Cercopithecidae.

Most of the G-bands of the human karyotype (2n = 46) easily match those in the karyotypes of macaques and baboons (2n = 42), in spite of the differences in the chromosome number.

They are, however, much more difficult to match in the karyotypes of the white-handed or common gibbon (*Hylobates lar*) to which humans nevertheless appear to be more closely related on the basis of the anatomical and molecular evidence.

The chromosomes of *Hylobate lar* virtually have no homology with other species of *Hylobates*. *Hylobate lar* (2n=44) and *Hylobate syndactylus* (siamang: 2n=50) are so similar genetically and anatomically that viable offsprings have been raised. Yet these hybrids have two quite different haploid sets of chromosomes mixed in each of their cell nuclei (2n=47).

TABLE 10.1
Rates of Chromosomal Evolution in Primates

• Fast	Hylobatinae (gibbons)
	Cercopithecinae (guenons and vervet monkeys)
	Aotinae (owl and titi monkeys)
• Medium	Ponginae and Hominidae (great apes and Man)
	Colobinae (leaf monkeys)
	Most of the Cebidae
• Slow	Cercopithecidae (macaques and baboons)
	Callithricidae (marmosets and tamarins)

Chromosomes in the gibbons seem to be evolving very rapidly (Table 10.1); the pace of evolution, however, is not commensurate with their anatomical or genetical rates of change. Most probably, their social structure and ecology (significant features being monogamy, territoriality and arboreality) assist in the incorporation of chromosomal variations through genetic drift. *Callicebus* (titi monkeys), which has a marked social and ecological resemblance with the gibbons, also reveals extensive chromosomal diversity. On the other hand, macaques and baboons, which live like unstable groups and have large local populations, reveal a much slower chromosomal change.

Most probably, the social structure and ecology is closely related to the rates of chromosomal evolution in primates. Other factors too might also be involved therein. For instance, this phenomenon does not readily explain chromosomal diversity in *Cercopithecus* (including guenons and vervet monkeys). In other words, the rates of chromosomal evolution vary widely across primates. These rates are often correlated with factors tending to promote random variations between populations.

A small number of chromosomal rearrangements are common to evolution and pathology. These are the intra-chromosomal changes (peri- and para-centric inversions). A large number of these changes observed in man can be considered as true reverse mutations. The fact that these same rearrangements are observed in evolution and pathology, and that when they are observed in pathology, no characteristic modification of the phenotype of the carriers can be observed, suggests that the role of these rearrangements in evolution is almost restricted to that of a gametic barrier.

Satellite DNA, which comprises about 4 per cent of the human genome, consists of short, simple DNA sequences repeated millions of times and localized to specific regions on all the chromosomes. Of the four major satellite DNAs in Man, three are also present in the genomes of the great apes. One human satellite II DNA seems to have been eliminated from the chromosomes of chimpanzees, but is present in the gorillas and orangutans. Satellite DNA, with some exceptions, appears to be evolving rapidly and is among the most obvious, quantitative and qualitative, alterations in the genomal DNA correlating with the evolution of species. Satellite A DNA of the chimpanzee shares many physical properties with human satellite III DNA. This and cross-hybridisation studies indicate underlying base sequence homology (Seth *et al.,* 1976). There exist chromatin similarities between man and chimpanzee (Seth, 1975).

There is a strong conservatism for chromosomes, which are carriers for Giemsa-11 because the composition of their satellite DNA has undergone relatively few changes over some 20 million years of evolution (Dutrillaux, 1979). However, when we consider more distantly related families, such as the Hylobatidae or the Cercopithecidae, the divergences may be more important, as revealed by the absence in them of Giemsa-11 staining and of human satellite DNA (Jones *et al.,* 1972).

Advances in DNA technology have provided us the information regarding the genetic relationships among primates. DNA hybridisation studies, in particular, attempt to provide divergence date(s) for the hominid line. Understanding of the intricate patterns of genetic diversity is possible today by conducting gene cloning, rapid DNA sequencing, restriction enzyme analysis and polymerase chain reactions. These patterns of diversity provide insights into population affinities and the evolutionary processes that created them. Molecular phylogenetics attempts to answer questions regarding how long our species has existed, who were the first *Homo sapiens*, where they lived and thereby facilitate understanding of human evolution.

The molecular intricacies of the genome are based on the definition of alleles as varying forms of a gene (i.e., a protein coding sequence) or as varying of homologous nucleotide sequences at specific loci (whether the loci contain coding sequences, pseudogenes, introns, restriction sites, tandem repeat sequences, etc). Evolutionary rates vary between different classes of alleles (Nei, 1987). Phylogenetic radiation from a common ancestor usually leaves a characteristic mark on allele frequency variations among the descendant populations (Cavalli-Sforza and Edwards, 1971; Cavalli-Sforza and Piazza, 1975). But gene flow between populations is not biologically prevented, as it is between species; population frequencies do not reveal both who and when of evolution (Long, 1993).

DNA sequence analysis and most analysis of mtDNA have proven quite ambiguous in attempting to determine the two closest relatives among humans, gorillas and chimpanzees. Probably, the preferred phylogeny relies on the choice of out-group and clustering techniques. A few regions of nDNA have been analysed and are similarly difficult to interpret. For instance, simply in the β-globin gene cluster on chromosome 11, the $\psi\eta$ sequence links humans to chimpanzees, γ^1 links gorillas to chimpanzees, γ^2 links gorillas to Man, and β-globin itself has linked chimpanzees to gorillas and Man in different studies.

Mitochondrial DNA and nuclear DNA provide convenient replacement hypothesis; the evolutionary rate (Table 10.1) of the former is relatively accelerated than the latter (mtDNA being maternally inherited). Brown (1980) using restriction enzymes to study mtDNA variation in 21 racially diverse individuals suggested that the time since divergence from the ancestral morph (allele) was between 18,000 and 360,000 years ago. Cann and co-workers (1987) estimated the mean rate of mtDNA divergence within humans lies between 2-4% my (four times higher than Brown's assumed rate) using high resolution restriction mapping. While the common ancestor of all surviving mtDNA types was estimated to exist between 140,000 and 290,000 years ago, the most ancient non-African branches of the tree arose between 90,000 and 180,000 years ago. Cann and his co-workers further inferred that Africa was a likely source for the entire human mtDNA gene pool, approx. 100,000 years ago. Other studies, using PCR, supported the findings obtained from the restriction sites (Vigilant *et al.,* 1989,1991; Vosberg, 1989). Foran and his co-workers (1988) estimated the coalescence time was 238,000 years based on comparisons of chimpanzee control regions and human sequences, and agreed with the restriction site estimates discussed above.

Allele frequency analyses of nDNA polymorphisms provide strong evidence for the Negroid and Caucasoid-Mongoloid split (Nei and Livshits, 1989) supporting the protein data. Multiregional evolution shows human DNA sequences (obtained from different origins) to be identical and that these differed from the normal chimpanzee allele by only three nucleotides. Humans and chimpanzees share many common alleles for the major histo-compatibility complex genes (Long, 1993). Bowcock and his co-workers (1991) suggest that *Homo sapiens sapiens* arose in Africa 100,000 years ago and subsequently expanded to other geographic regions, replacing the existing populations such as the *Neanderthals.*

The primary division between African and non-African populations seen in DNA – both mitochondrial and nuclear – analyses is more consistent with the replacement than the multiregional evolution. Efforts to reconstruct human demographic history have overshadowed our pursuits to understand and explain the causes of evolutionary variation – i.e., the roles of selection, mutation and genetic drift (Livingstone, 1991). The theory of punctuated equilibrium holds that significant evolutionary change only arises coincident with branching speciation events (Gould, 1982). These events are instantaneous in geological time, and aside from mild morphological fluctuations, the major features of most species are evolutionary static. In contrast, Wright's (1931, 1982) view holds that the adaptive evolution is most likely in a widespread species subdivided into local demes with some gene flow. In other words, evolution occurs in occasional rapid events that correspond to ecological changes and not in speciation events.

Gene Mapping

Based on comparative gene mapping, Turleau and co-workers (1983) strongly suggest that the Hylobatidae diverged from the common stem leading to the Pongidae after the Cercopithecidae had diverged. The Pongidae, again, are the most studied family, and within it, the chimpanzees are the most studied species. Generally, workers have consensus on the localization of genes found in the species investigated. The localization of 27 genes is identical in the same chromosomes of man and chimpanzee, that of 22 genes in the same chromosomes of gorilla and man, and that of 14 genes in the same chromosomes of orangutan and man.

By and large, the results obtained by gene mapping and by chromosome banding assays are very similar. Therefore, the notion of very strong conservatism of the genetic material appears very well established. The conservatism of the X-chromosomes, thus, could be a general rule for the whole karyotype. It might be stronger for the X-chromosome, or merely easier to demonstrate for this chromosome than for the autosomes.

None of the primates appear in the fossil records earlier than about 65 mya, and those of modern aspects about 56 mya. As such, the maximum divergence times amongst living primates may not exceed 60 million years or so. There is about 35 per cent change along the albumin lineages of the simian primates since their separation from those leading to living prosimians, or about 0.6 per cent immunological differentiation for each million years per lineage. For instance, two typical primates whose albumins differ by 12 per cent (i.e., 6 per cent change along each lineage) will have last shared a common ancestor 10 million years ago. If this is true, then gorilla, chimpanzee and Man whose albumins differ by about 5 per cent from each other, separated from one another about 4 million years ago. This would be a minimum estimate if the radiation of the primates began more than 60 million years ago.

Immunology suggests that molecular evolution had slowed in the lineage leading to Man. Human serum proteins show only small antigenic differences from orangutans and gibbons, and negligible differences from gorillas and chimpanzees. Considering the extensive morphological evolution involved, the extent of protein evolution between humans and other hominoids is surprisingly small. Alternatively, if the 'molecular clock' is applied to immunological distances and the divergence between the hominoids and the Old World Monkeys arbitrarily at 30 million years ago, the human-African ape split occurs at 5 million years ago. Protein evolution could have slowed in the descent of hominoids, but this would not rule out a relatively late common ancestry of Man, chimpanzee and gorilla.

11

Palaeodemography

Palaeodemography is the specific study of past human vital statistics and not merely a general study of the prehistoric populations; it is more than just estimating numbers. Although interest in aboriginal populations' size dates back hundreds of years, reliable estimates remain elusive.

The archaeological approaches utilise data on settlement patterns, size and number of houses and refuse areas within the settlements, and faunal analysis. On the other hand, the biological analysis of the archaeologically recovered human remains contributes data on the life expectancy, death rates, and population size directly when certain variables can be controlled (Ubelaker, 1974). A human burial contains more anthropological information per cubic meter of deposit than any other type of archaeological feature. A burial represents the latent images of a biological and cultural person frozen in a distinctly delimited segment of space and time (Peebles, 1977). All these archaeological approaches are vulnerable to sampling problems. The most useful approach, although additional, in establishing the upper limit of population size is the estimation of ecological resource potential from the environmental and cultural data.

Recent research in palaeodemography focuses investigation of not only the population size but also of such variables as village size, number of occupants per house, family size, population composition (ratios of males to females or of adults to subadults), rates of mortality and fertility, and population growth Much of this research overlaps areas of prehistoric nutrition, subsistence, epidemiology, and population dynamics.

Palaeodemography deals with a composite field where forensic anthropology, archaeology, demography and physical anthropology interface. An increasingly important aspect in these studies is the documentation of the methodology employed in determining the age and sex. There also exists an urgent need for more reliable demographic data from different geographical areas and different chronological periods to better document prehistoric demographic variation. Data must establish that the recovered skeletal material is representative of the true distribution of deaths in a population. The reality is that we know little about variability in palaeodemography. Few studies have been published and many of these are of little value since they are based upon poorly documented samples or utilized unreliable or substandard techniques.

Today, through osteological analysis, our knowledge of previous cultures and human populations is greatly increasing. Information and or the data thus acquired has historical as well as medical significance to modern humans. Therefore, the studies of the human bone are invaluable to explaining and understanding the past. From the study of bone, one can understand the nutritional status, diet, disease processes, factors affecting mortality and life expectancy, biological responses to environmental stressors and aging of our ancestors. Also thus gained is the wealth of information about the prehistoric cultural practices and social structure, population movements and about tribal origins.

Demographic analysis of the age and sex distribution of the skeletal series provides statistics regarding the age-specific mortality rates, age distribution within the population, longevity, sex ratios, and natural rates of increase or decrease (growth rates). The regional increase in the population size and density is generally estimated by the size of the site and its distribution, number of houses, and ethnographic analogy. Analytical studies of the demographic structure of the populations help define the trends in human longevity, the effects (the environment and the changes in the environment) had on the human life span, and provide insights into the underlying causes of population change. For instance, knowledge of the mortality rates, fertility and sex ratios help explain how well a population was adapted to its cultural and physical environment, why it grew in size or declined towards extinction. Population estimates are attempted in utilizing the crude mortality rates (obtained from the life tables), the length of time represented by each ossuary (obtained from the archaeological data), and the total numbers of individuals in the ossuaries. Comparison of, both the archaeological and the ethnohistorical data against the 'resulting population size estimates, suggests the nature of the socio-political unit serviced by the ossuaries. The recovery of human remains, however, during traditional excavation in the areas provides little assurance that the materials recovered are adequate samples of the populations. Palaeodemography also provides insights into why certain cultural practices may have developed and continued.

Settled villages rather than individual homesteads mark the Neolithic, and the metal ages include urban centers with satellite peasant villages in Iran and Iraq (Rathbun, 1984). This pattern of land use is still observable with contemporary groups.

Demographic attributes and their analysis require consideration of a variety of social and biological variables. Vital statistics of birth and death are frequently used as measures of quality of life in the contemporary societies.

Methodology

The adaptation of demographic methods to palaeodemographic study must be accompanied by an explicit appreciation of assumptions made about the uniformity of the biological processes in our species – in the past and in the present – as well as the biases introduced by the nature of the archaeological record. We must also realize that our estimates of rates and numbers are just that – estimates. It must also be recognized that the demographic information cannot specify the actual content of the past diets. Instead, palaeodemographic study indirectly defines the impact of life ways upon key demographic parameters such as life span, fertility, group size, population density, and population increase.

Some palaeodemographic features are easier to estimate than the others. With appropriate techniques, mortality patterns can be estimated from an analysis of the skeletal remains. Fertility is quite difficult to measure although it is possible that age of menarche can be posited. The mean number of children born to each female is still a desirable but elusive parameter, as is the rate of population increase. And in certain contexts, group size and population density may be inferred, within limits.

The ethnohistorical and archaeological information suggests that the tidewater Potomac ossuaries, in U.S.A., contain nearly a complete representation of all individuals who died in the contributing populations during a culturally prescribed number of years. The mortality curves show the highest frequency of deaths in the first five years of life, a dramatic decline through childhood into adolescence, a steady increase during early childhood to a maximum adult death frequency between 30 and 35 years, followed by a steady decline after 35 years of age. Differences in mortality between populations are best demonstrated by survivorship curves. For instance, the curve of survivorship accentuates the differences between these two populations and shows that Ossuary II Population has a higher infant and adolescent mortality, but greater adult longevity (Ubelaker, 1974).

Comparisons of life tables for the two populations reveals differences in age-specific mortality; the probability of dying in Population I is greater than that in Population II for all ages except 15-20 (adolescents) and 20-25 (young adults) years. Life expectancy for Population II was greater than that for Population I in all age intervals; crude death rate for the Ist Population is 47.8 and that for the IInd is 43.5 per thousand. Comparing these death rates with some other skeletal populations show that these figures are lower compared to South Dakota, Greece and the Kentucky populations and are higher than the death rates for the Ohio and Ontario series. Data on life expectancy and age-specific mortality provides an evaluation of the overall population fitness and health status. Age and sex-specific mortality estimates provide the basic data for the various forms of palaeodemographic inferences.

Demography of extinct human groups is usually based on the assumption related to the rates of ontogenetic development and aging in human populations. For instance, is it proper to assume that the ancient populations aged at the same rate as individuals in the modern populations whose standards are used for projection into the past? There is evidence that other species of animal's age at different rates. The great apes, our nearest relatives, have a life span appreciably shorter than our own. Therefore, in order to make such comparisons, we must assume that the biological potential for equivalent tempos of development were nearly identical.

Studies of hunter-gatherers in contrasting environments point to the equivalent levels of fertility and mortality, levels that are principally affected by subsistence strategy and pathogen load rather than different genetic heritage (Howell, 1976). There is evidence that at least one skeletal indicator sometimes used to estimate age, cranial suture closure, appears to display a secular trend towards earlier maturation in certain recent populations (Masset, 1971). As there is evidence that individuals age at different ages, it is advisable to compare groups of relatively large size – thereby attempting to compensate for individual variation – and to use comparable developmental stages as standards for interpopulation comparisons (Buikstra and Mielke, 1985). These basic assumptions are viable for living populations but not for the past skeletal groups of fossil hominids. It has been argued, based upon observations of living groups and simulation studies (Weis, 1973, 1975), that stability is introduced in small groups relatively early and rapidly after significant deviations. Thus, a cemetery should provide a reasonable accurate means of characterizing the vital rates of a population.

Recent criticism of palaeodemographic methods has centered on biases introduced by nature of reference samples and the population-specificity of techniques for estimating age in the skeletal remains (Buikstra and Königsberg, 1985). Considerable error is introduced because of the method followed in the analysis. For instance, to estimate the age of the child at death, usually one applies methods generated from model samples; this is an extension of the principle of uniformitarianism. It is based on the assumption that the biological process of aging is the same today as in the past and, therefore, the modern data and its observations can be extended to the prehistoric skeletons.

Although considerable interest exists in anthropology for demographic analysis of past populations, broadly speaking, the database is insufficient for a viable temporal comparison of survivorship and life expectancy. Empirical and model life tables have been often generated from the available skeletal remains. However, problems of sampling, under-representation of preadults, unsystematic excavations, and sporadic osteological analysis makes demographic specifics quite tenuous.

Sexing

Accurate estimation of sex and age at death of the skeletal remains is especially important in the development of demographic reconstructions. The practice of sexing remains through assessment of associated grave goods should be avoided. Similarly, to sex skeletal remains when accompanied by an infant or a young child as a female would be erroneous for certain prehistoric populations placed juveniles with adult males (Buikstra, 1972). Sexing of the skeletal remains usually is based upon morphological observations and metric determinations. Sex identification often can be determined by the differences in the pelvis and skull; even bone fragments can be sexed, for some chemical components of the bone differ between men and women. It is mostly emphasized that the sub-pubic angles of the female pelvis tend to be relatively U-shaped and broad compared to the V-shaped and narrow pelvis of the males. Though the skull shows dimorphism, yet these characteristics are less explicitly tied to sex differences in growth and development than features of the pelvis. Phenice's (1969) standards of sexing involve structures in the pubic region that are anatomically related to the attachment of the crus penis and the crus clitoris: ventral arc, the sub-pubic concavity, and a ridge on the medial aspect of the ramus, listed in decreasing order of reliability. The major advantage of this technique is that it can be applied even to the fragmentary and fragile specimens.

Metric observations on sex determination are based both on the univariate and multivariate statistical techniques. The development of statistical standards represents an advance beyond the more subjective morphological observations, both in terms of technical accuracy and as a means of analysing fragments and isolated long bones.

Sexual dimorphism in the juvenile skeletons, though long felt, is usually based upon observations of dimorphism in the dentition. Significant dimorphism is reported in the canines (Garn, 1967; Lunt, 1969): enamel thickness (Moss and Moss-Salentijn, 1977); cemento-enamel junction dimensions (Barker, 1973); crown diameters (Owsley, 1981); and measurements of the tooth roots (Garn *et al.,* 1979). It follows that sex diagnosis based upon dental attributes is advantageous in skeletal series with marginal preservation. Dental studies would thereby provide means for learning about palaeodemography in regions where imperfect preservation has precluded such work in the past.

Ageing

No single skeletal indicator of age at death is ever likely to accurately reflect the many factors, which accumulate with chronological age, each of which can contribute valuable information to the age estimate. Age at the time of death can be estimated very closely by observing the teeth and the degree of fusion between different parts of the same bone, especially for children and young adults. However, for older people, the estimates are less exact and rely more on changes in the joint surfaces, fusion between skull bones and microscopic details of the internal bone structure. An indicator which both significantly reflects biological age and whose information content is independent of other indicators will be useful to arrive at a final age estimate; whether under forensic or archaeological conditions. Accurate demographic reconstruction demands reliable age estimates at death for all the individuals in the skeletal population. The major aspects of determining/inferring the age of the fossil hominid remains is the development, the eruption and the attrition of the teeth. These characteristics, however, must be used with utmost circumspection.

- *Subadult ages:* derived from dental formation and eruption and from maximum femoral lengths.
- *Adult ages:* calculated from examinations of symphyseal facies of the pubes and the degree of microscopic cortical remodelling in the femora.

Histological methods for estimating the age of skeletal remains, especially useful in physical anthropology, provide greater accuracy of age estimation for adult skeletal remains than do traditional gross morphological methods (Stout and Gehlert, 1980); these potential methods of superior accuracy can also be applied to fragmentary remains (Stout, 1988). Averaging age estimates by Kerley's (1965) six formulas (mean regression) provides the overall greatest accuracy and reliability.

Too little consideration has been given to the differences between the two sexes in the ageing processes. If age estimations are carried out using a large number of age markers, over and above the individual cases of a pathological nature, one can correctly estimate the physiological age of the skeletal remains examined. These would be correctly placed on a relative scale which would be nearer to the complete one, the more age markers used and the more precise the techniques of analysis. In any case, comparisons can be made between sub sample of the same population, between diachronic populations – from the same cemetery – and between populations, which live in the same area. In fact, over and above the complex mathematical formulae, whenever "age-related events and processes affecting skeletal remains are examined independently of age estimation, i.e., demographically, their patterns of occurrence make sense in light of modern skeletal biology" (van Gerven and Armelagos, 1983).

In the age distribution of adults and juveniles, account should be taken of the stadium of development of the synostosis between the epiphysis and the diathesis of the long bones and the degree of ossification of the girdle and the axial skeleton (Ferembach *et al.,* 1977-79).

There are two schools of thought concerning adult age estimation in skeletal remains: a multivariate approach which employs multiple indicators like endocranial suture closure, dental attrition, involution of trabecular structure, metamorphosis of the symphyseal face of os pubis, and arthritis development (Acsadi and Nemeskeri, 1970; Sjovold, 1975); and the single key attributes such as morphology of the pubic symphysis (Meindl and Lovejoy, 1983; Buikstra and Mielke, 1985).

Dental maturation, particularly mineralization, is less vulnerable to environmental stresses than other aspects of cranial and post-cranial development. Thus, standards for estimating age-at-death from the dentition are considered to be maximally reliable. The accuracy of the vastly preferable Ubelaker's (1978) chart of dental development, particularly for older juveniles could also be enhanced through the establishment of reliable methods for sexing immature skeletal remains. Usually, there are no consistent differences between sexes in the emergence timing for the deciduous dentition (Tanner, 1962). However, it is well-established that in the permanent dentition of the females, with particular reference to those teeth with latest emergence, maturation tends to take place significantly in females than in the males (Buikstra and Mielke, 1985). Therefore, establishing the sex of a skeleton first is likely to increase the accuracy of age estimates in juvenile remains.

Under certain circumstances, age derived from the femur length may provide more reliable estimates than dental eruption (Ubelaker, 1974, 1978; Hoffman, 1979). It is, however, preferable to simply compare lengths per stage of dental maturation without introducing the intervening concept of chronological age.

Yet another method of estimating age, with an excellent potential in palaeodemography, based upon the progressive changes upon the face of the auricular surface of the innominate has been established; the dorsal aspect of the innominate being much less fragile than the ventral region, which bears the pubic symphysis (Meindl *et al.,* 1980). Despite many limitations, multiple age indicators combined statistically are favoured (Lovejoy *et al.,* 1977; Acsadi and Nemeskeri, 1970; Sjovold, 1975; Buikstra and Mielke, 1985).

Results obtained from aspartic acid racemization analysis are consistent with those based upon other micro- and macroscopic osteological age estimates of ancient skeletal remains (Masters and Zimmerman, 1978). The technique based upon the accumulation of D-aspartyle residues in the tooth enamel or dentine (protein fraction) is however limited to well dated contexts and those where internment has been in a cool and stable climate; the racemization rate is markedly affected by temperature changes.

Kerley and Ubelaker (1978) propose microscopic observation of the osteon system (counting of remodeled osteons and their comparisons to a standard) in the cortical bone (long bone sections) in adult remains. This technique is rapidly gaining popularity. Recent advances in age estimation, therefore, include both the refinement of existing methods and the development of new strategies.

Comprehensive studies of palaeodemographic factors of population size, density, mortality, survivorship, and life expectancy based on skeletal remains in south Asia are few at the present time (Kennedy, 1993). It has been reported that the period of the post-Pleistocene hunting-gathering life ways brought with it a dramatic population expansion following the longer and more stable Palaeolithic period (Fairservis, 1975). The apparent stability of the food resource base for many may be reflected in the relatively high ages at the time of death for the skeletalized individuals from south Asia compared to the lower ages at the time of death for the representatives of the later farming-hunting communities. At the terminal Pleistocene site of Mahadaha in the Gangetic Plain, some human remains have been assessed as having died at their 50s and 60s years of age (Kennedy, 1984).

Stature and body size decreased in many parts of south Asia with the abandonment of a full-time hunting-gathering economy, and it is interesting that individuals who presumably died in their fifth or later decades are seldom observed in the sedentary and agricultural communities. Aged individuals appear in the terminal Pleistocene site of Mahadaha. Studies on Sudanese in Nubia suggest that significant population growth occurred with the transition from gathering and hunting to agriculture (Martin *et al.*, 1984).

Parity and Weaning Age

Accurate estimates of fecundity and fertility in the past populations would not only enhance many palaeodemographic aspects like number of pregnancies, age of first reproduction, length of the reproductive period etc but would also help infer relative disease impact upon demographic patterns, provide data on prehistoric human ecology both for reconstructing local adaptations and for modeling evolutionary change (Buikstra and Mielke, 1985).

Relationship between pelvic stresses of childbearing (e.g., remodelling of bony structures adjacent to pelvic joints) and skeletal features is well known (Putschar, 1931). There appears to be a statistical association between the number of reported (full term) pregnancies and the degree of pubic remodelling, especially the pitting of the dorsal face. Fertility and reproductive patterns can also be studied using skeletal indicators of stress and biochemical evidence of diet as estimates of weaning age. For instance, early weaning (pointer to frequent pregnancies) would result in elevated (frequencies of) non-specific stress indicators at increasingly younger ages in juveniles (Buikstra and Cook, 1980). Whereas, trace element patterns would provide evidence of dietary supplementation (Sillen and Smith, 1983). There is evidence of close linkage between length of breast-feeding and post partum amenorrhea; this has strong implications for fertility and population increase (Lesthaeghe and Page, 1980).

Population Size Estimates

Community size and regional population density are the key variables for the understanding of human adaptation and evolutionary change. Estimates like site size and density, dwelling size and number, artefact number of human skeletons, together with burial mound types based upon shape, size and location help construct models of specific gradual regional population increase.

Professional estimates of the size of the New World aboriginal population at the time of initial European contact display remarkable variability. For instance, hemispheric estimates of the prehistoric populations range from 8.4 m to about 100 m; estimates for America North of Mexico range from 900,000 to nearly 10 m. These estimates suggest that the population densities vary from a relatively sparse occupation of two persons for every 10 sq kms to a relatively dense population of 21 persons for every 10 sq kms. Additional figures vary widely between these ranges, reflecting both the limited database and the subjective, varied methodology employed in their derivation. Recent population estimates of the prehistoric North American tribals are markedly higher than those published earlier (Mooney, 1928; Kroeber, 1939). Although these estimates reflect the current trend among prehistorians to accept larger prehistoric population estimates than suggested by Mooney; they fall appreciably short of those predicted by Borah (1964) and Dobyns (1966).

Mortality Patterns

Palaeodemographers often use computed mortality rates as indicators of the quality of life in extinct human groups to evaluate relative successes under different subsistence periods; Life Table parameters forming the primary mode of analysis. Deviations from normal mortality patterns would reflect impact of special causes such as epidemic diseases or massacres of the populations.

This data helps establish population health status, provides unique information about social complexity of the past, and estimates of demography (like population size, age and fertility) enable us to relate this to the variety of life ways characteristic of living human groups.

Growth rates and nutritional status can be measured by examining bones; for they show many indicators of disease occurrence and nutritional stress. This has direct relevance to modern populations. For instance, it is well known that the incidence of bone cancer in present day Tennessee, Alabama, and Georgia, in and around the Tennessee River Valley, is among the highest in the United States. By studying the skeletal remains of earlier populations of this area, one can determine whether they showed a similar pattern or not. This would help answer the antiquity of the disease and its causes. Such studies are relevant to the health of the modern populations.

The skeletal remains provide one with information needed to evaluate how the populations lived and died. The adaptive interrelations of the biological and cultural aspects provide one of the most advanced approaches characterizing present day anthropology (Armelagos *et al.,* 1981).

Numerous demographic studies on Nubian populations have revealed the potential for understanding the forms of biocultural adaptation of the past populations living in that area (Coppa, 1988); biological stress indicators were used to evaluate the impact of social factors on ancient populations (Martin *et. al.,* 1984). Data on their demographic structure shows:

- a diminution in sexual dimorphism in recent times in contrast to the earlier periods;
- female mandibles are characterized by large dimensions, although they were smaller than the males;
- the maxillo-facial complex has undergone a considerable reduction;
- a definite underestimation of the mortality of the infant population;
- the juvenile and adult data points to a certain reduction in the death rate in recent periods;
- that the transition from the Meroitic period to the Christian period is marked by a change in the mortality models; and
- in the Christian population, more than 25 per cent of the deaths are concentrated in the 20-24 years age group; whereas there is higher

mortality in the older age-groups among the Meroitics except the over 50 years group, where it is much higher for the Christians.

Economic and cultural changes are powerful determinants of the patterns of morbidity, mortality and stress. A mortality rate is evidenced in the Indian Knoll population (Ohio, USA) in infants than in the toddlers, a pattern like that of modern urban populations but unlike that of many modern peasant populations (Cassidy, 1984). The relatively high mortality rate in the newborns may indicate some social mortality (infanticide). About 5 per cent of the population survived past the age of 50 years. Life expectancies were lower for females than males at all ages. Perzigian and co-workers (1984) observed a noticeable decline in the dental and skeletal health in the Ohio River Valley, which probably reflects a growing synergism among diet, population density, and disease.

The palaeodemographic data from the Central Valley indicates that the overall subadult survivorship increased through time while adult survivorship decreased; through time, birth spacing decreased as population grew, and there was increased mortality of young adult females related to increased exposure to the risks of child-birth (Dickel *et al.,* 1984). The types of morbidity and mortality culturally perceived as manipulable by changing subsistence patterns might be a factor in subsistence shifts. Multiple stress indicators are used to determine the degree and pattern of stress in the populations.

Though the Upper Palaeolithic Ein Gev female burials and the Aurignacian remains at El Wad and Kebara provide little information on palaeodemography but do indicate differences in the burial practices that are repeated in the Natufian (Smith *et al.,* 1984). At the Natufian and Neolithic sites, mortality is normally highest in the first year of life, falls slightly in early childhood, and is the lowest in late childhood and adolescence. In populations with low life expectancy, the percentage of infant deaths may exceed 40 per cent. This is associated with a high frequency of young adult deaths with only a few individuals surviving to middle or old age. Increased longevity thereat is associated with both a lower frequency of infant deaths and a higher ratio of older to younger individuals. Analysis of the ratio of young to old adults shows that longevity increased in the Neolithic period and showed a second but temporary increase in the late Hellenistic and Roman period. Arensberg (1973) estimated an average life span of 30 years for the Epipalaeolithic, increasing to 34 years in the Neolithic, 39 years in the Hellenistic and Roman periods, and dropping slightly thereafter to 32 years in early Arabs. Female life expectancy over the same periods averaged some 4 years less in each group.

Growth

Growth curves of long bone development are good indicators of nutritional stress related to subsistence. Whereas adult stature serves as a fairly poor indicator of nutritional stress since growth is controlled and influenced by a multitude of factors.

Disease, Diet and Demography

Studies of prehistoric populations and their burials reveal the type of interaction with their environment and how they in turn were affected by changes in the physical world they occupy. These studies, especially demography and disease patterns help understand the ecology of palaeogroups, show effects of changes in the ecology and social and economic patterns on their health and life span, and facilitate assessment of health cost of agriculture, and the origins of some modern human diseases.

Many contemporary diseases, f.i. osteoporosis, tuberculosis, etc, have been reported in earlier human skeletons. Interestingly, osteoporosis is common in the living and prehistoric Eskimos of both sexes.

Studies of earlier populations reveal that high levels of tooth decay are typical only of agricultural groups. Perhaps sticky carbohydrates, common to most agricultural diets, have something to do with the epidemic of tooth decay modern populations are experiencing.

Tuberculosis has been found in skeletons as early as 5000 years BP in the Old World and by at least AD 1000 in the New World. It has been found to be associated with keeping livestock and living in sedentary or urban centers. A curious relationship between tuberculosis and leprosy has been found. Skeletons rarely show signs of both diseases; as tuberculosis became more common in Europe in the late Middle ages, signs of leprosy in European skeletons declined. It is speculated that exposure to tuberculosis provides some immunity to leprosy.

Australopithecus

The relatively abundant evidence of *Australopithecus afarensis* from the Afar and Laetoli (Stringer, 1984), which dates between ca. 4 million years and 3 million years, has clearly demonstrated that the fundamental human adaptation of bipedalism appears to predate the first clear evidence of tool making and expansion of the brain beyond the level found in our nearest relatives, the African apes. The phylogeny favoured by Johanson and White (1979) would displace the gracile Australopithecine, *Australopithecus africanus*, from the position ancestral to *Homo*, which it generally occupies in most accounts of the hominid evolution. Instead it is "relegated" to an ancestor of the robust australopithecines (*Australopithecus robustus* + *Australopithecus boisei* = *Paranthropus* of some) only. *Australopithecus afarensis* instead represents the common ancestor of all subsequent hominids according to Johanson and White (1979). Although if it is equally related to the *Homo* and *Australopithecus* lineages, it is unclear why it should be classified with *Australopithecus*. Deviation in the gnathic morphology of *Australopithecus africanus*, a deviation culminating in the morphology of the robust australopithecines, eliminates it from the lineage of *Homo*.

Homo habilis

The prevailing interpretation of the archaeological levels from Olduvai Bed I is that they represent living areas, or "home bases", of early hominids, specifically *Homo habilis* (Potts, 1986). The home base interpretation arises from the presumed similarity between the accumulations of stone tools and bones from archaeological sites and the composites of modern, tropical hunter-gatherers. Hominids are thought to have brought animal bones to sites that were also the focus of social activity. Animal foods were shared and tools were used to process animal tissues and for other tasks typically carried out by hunter-gatherers at home bases. Olduvai hominids used their sites over a longer period and in ways different than those practiced by the recent hunter-gatherers (Potts, 1984).

Neanderthals

Coon (1963) proposed that their large projecting noses, maxillary sinuses and infraorbital foramen all reflected adaptation to a periglacial environment. Added to this list are the cranial characters like frontal sinus development and degree of occipital bunning, which could be, so explained (Brose and Wolpoff, 1971). Coon also believed that their body proportions were like those of modern cold-adapted people.

For most fossil hominids, the typical nasal morphology is of a flat and fairly broad nasal opening, as would be expected for relatively moist and warm environments at low altitudes. But the *Neanderthal* skulls from Eurasia are unusual for their extremely voluminous and projecting nasal openings. This form of nose could be explained through adaptation to conditions of low humidity and temperature. Since their nose was also broad rather than narrow, there would appear to be no obvious analogy to it among the recent populations. No present day populations are adapted to an environment, which closely resembles that of Europe during the last glaciation, where there may have been unique combinations of temperature and humidity.

Neanderthal skeletons are also unusual in their limb segment proportions in that the distal limb segments (radius, ulna, tibia, and fibula) are short in comparison with the proximal segments (humerus, and femur). Such limb segment differences are indicative of selection operating to minimize surface areas of extremities, or of repeated vaso-constriction showing growth, in a cold environment (Coon, 1963; Stringer, 1984). This distinctiveness of early anatomically modern limb proportions is a reflection of gene flow or population movement from warmer environments at the transition to "modern" *Homo sapiens* in Eurasia (Trinkaus, 1980).

12

Palaeopathology

The study of palaeoanthropology, one of the oldest fields of human biology, relates to extrapolation from a few bones to a population. For instance, human skeletal remains have been often used for studies of life expectation in ancient societies, the existence of cemeteries being a reasonably accurate reflection of their mortality patterns. Broadly speaking, extensive archaeological investigations into the domestication and urbanization processes have been made that little is known about the physical aspects of the groups in general and their pathology in particular. The major problem of palaeotaxonomy arises from the fact that the species loses its clear-cut reality when considered in the dimension of time.

Though soil type seems to have little influence on early decomposition of bone, there is no doubt that different soils have different effects in the long-term preservation of bone (Cornwall, 1956; Brothwell, 1963a). Major pathological changes in the bone may be simulated by erosion due to rodent teeth, beetles, plant roots, high winds, etc. Some relatively rare diseases, but of characteristic appearance in bone, may be confidently diagnosed than many common lesions. Many other killing diseases leave no imprint on the bones. The commonest disease to be found in the ancient bones is arthritis (Moodie, 1923; Zorab, 1961; Sandison, 1968). Osteoarthritic changes have been reported in *Neanderthal* man from Shanidar (Goldstein, 1963), *La-Chapelle-aux-Saints* and *Cro-Magnon* man (Brothwell, 1963a). The other most common change in the ancient bones is non-specific inflammation (Sandison, 1968; Brothwell, 1963b).

Palaeopathology, devoted to the study and diagnosis of diseases in ancient human remains, provides medical insights on the history and ecology of modern human diseases. For instance, childhood illness or malnutrition can be detected by abnormalities in tooth enamel and bone mineralization.

Many palaeopathologists have described evidence of infection around the middle ear and air sinuses in many cultures. *Rhodesian* man from Broken Hill in Zambia seems to have both dental and ear infections (Sandison, 1968). Specific infections like TB are well documented by the ancient remains. There exist skeletal evidences for leprosy, syphilis, poliomyelitis, variety of tumours, osteosarcoma, and carcinoma etc (Brothwell, 1967; Sandison, 1968).

Lengyel (1968) reports the existence of a correlation between:

- sex and citrate concentration;
- individual's biological age and the amount of calcium carbonate and collagen contained in the vertebrae; and
- the degree of bone composition and its chemical and micro-morphological structure.

Palaeopathology is an important discipline for the scientific study of man as it provides:

- background information relevant to palaeodemography;
- material evidence, valuable for historic as well as prehistoric times, of a varying disease spectrum in the past;
- evidence of heritable variation, in respect of congenital anomalies due to a genetic basis, ranging from possible single gene differences to chromosomal aberrations;
- the frequency of traumatic injury and disease, degenerative joint diseases;
- helps us answer questions regarding the antiquity of the disease and its causes. For instance, the study of the skeletal remains of the earlier populations of the area help us compare the incidence of the disease in that area today, e.g. bone cancer in present day Tennessee, Alabama and Georgia;
- indications of diseases like tuberculosis, healed fractures, arthritis, and rickets;
- not only informs us about the history of the human disease, but can also occasionally help elucidate the causes of certain diseases, as well; and
- age progression of dental decay.

It was combined with palaeodemography to obtain a population perspective on health as equilibrium with disease.

As anthropologists, our main interest is in human well being, which is a balance of mankind with disease parasites and environmental stresses. On the other hand, palaeonutrition has two aspects: response of bone and teeth, and the trace element record in the bone.

Palaeopathological studies gained great impetus in the second half of the 19th century, the period of Broca and Virchow. The pathology of a society reflects its general conditions and growth and offers, therefore, valuable clues to an understanding of the total society.

The first healed fractures of animals as evidenced by callus are found in Permian reptiles. A healed fracture was also found in the *Neanderthal* man, *Cro-Magnon* man, the European Neolithic, and the American pre-Columbian, especially in Peru (Ackerknecht, 1962). The latter prevalence might be due to the use of special weapons like the maces. Numerous arrowheads have been found in the European Neolithic and in pre-Columbian America, imbedded especially in the vertebrae and extremities. Periostitis ossificans (inflammation) was first reported in Permian reptiles, later in the early mammals and eventually in the Neolithic man. Numerous prehistoric bone lesions have been described as syphilitic; they (bones) being morphologically stronger.

Evidence of dwarfism, achondroplastic and cretinistic, has been found in Egypt – in the skeletons and statues. Congenital anomalies of the vertebral column (spina bifida) and of the sternum have been reported from the Neolithic of Europe and Peru.

Next to traumatism, arthritis is the oldest and most widespread pathological lesion reported in palaeopathology. Among the hominids, arthritis has been observed all over since *Neanderthal* man. Chronic arthritis of the hip joint is reported in *La-Chapelle-aux-Saints*, temporo-maxillary arthritis in *Krapina* Man (*Neanderthal* Man), Melanesians, pre-Columbians and in New Caledonia. Numerous other joints are affected in *Neanderthals,* Upper Palaeolithic men (*Cro -Magnon*), Neolithic Europeans and the Pre-Columbian Amerindians.

Spondylitis of the cervico-dorsal and the lumbar region is seen in the *La-Chapelle-aux-Saints, Cro-Magnon*, European Neolithic, early Egyptian and the pre-Columbian American man.

A characteristic feature of spondylitis is the changing seat of lesion along the vertebral column. For instance, in early man and the primitives, lumbar spondylitis is frequent, whereas dorsal and cervical are rare. In modern man, involvement of this is often found. These localizations of the pathological areas indicate places of stress and strain as also the living conditions and posture of the individuals under reference.

Rickets and bone tumours in prehistoric men are generally rare compared to those found in the present day men. Dental pathology was first reported amongst the *Neanderthals* (pyorrhea) and then among the Neolithic, especially in Scandinavia.

Only a small number of diseases leave their marks on the bones. However, in mummies, the soft parts are preserved to a certain extent. Arteriosclerosis has been observed in the Peruvian mummies, while pleurisy, congenital atrophy of the kidneys, gallstones and liver cirrhosis etc have been demonstrated in the Egyptian mummies. The human skeletal and mummified remains from precontact cultures of the New World provide more direct evidence of disease. Unfortunately, only a few or relatively small number of diseases that affect humans leave any trace of their presence in the skeletons, and some diseases affect bones in similar ways, making an accurate diagnosis difficult. Diseases that affect the skeleton are most commonly long-term ailments such as arthritis, chronic infections, and certain dietary deficiencies. Most illnesses that kill people quickly, particularly acute infectious diseases such as influenza, small pox, typhoid, and measles, do not leave any of their marks on the skeleton. Palaeopathologists are, therefore, limited to a narrow field of inquiry.

Human skeletal remains of 24 individuals from complex mortuaries (ca 900 B.C. to 200 B.C.) in La Libertad, Guayas Province, Ecuador show artificial cranial deformation, relatively high frequency of infectious diseases, relatively low frequency of carious lesions (4 per cent), and no signs of trauma or porotic hyperostosis. This suggests that the population had a relatively rapid rate of dental attrition, which produced dentin exposure before the age of 20 and loss of most of the crown by the age of 50 years (Ubelaker, 1988). The estimated mean living statures of these was 153 cm for females and 170 cm for males; the adult mean age at death of this population was about 35 years with some individuals having survived till 50 years. The prehistoric Ecuador sample displays complex biological variation and evolution of prehistoric inhabitants, possessed general skeletal biology, facilitates synthetic review of biological temporal trends, and suggests a general temporal increase in certain pathological conditions and other indicators of population stress.

Recent studies of skeletal and dental stress indicators in the New World remains have shown that health problems were common. Furthermore, morbidity seems to have increased through time in many areas of the New World with increased dependence on agriculture. For instance, studies of human remains recovered from archaeological excavations in Ecuador document that 8,000 years ago; populations who lived in coastal Ecuador had low levels of infectious disease, anemia, dental caries, and various measures of nutritional stress. Samples of human remains from more recent time periods in Ecuador show regular temporal increases in the frequencies of these problems, apparently resulting from a more sedentary way of life and a less varied diet. As populations grew larger, and with a shift in subsistence began to live in more permanent settlements, sanitation problems inevitably developed, leading to an increase in infectious disease. There are increasing number of cases highly suggestive of treponemal infections in pre-Columbian New World skeletal remains: yaws, endemic syphilis, venereal syphilis, etc.

Population densities probably never reached levels necessary to sustain and spread many of the infectious diseases, such as influenza, typhoid, measles, smallpox, that were common ailments in Europe and West Africa in the late 15th century. Studies of human remains are invaluable for explaining and understanding the human past in as much as the nutritional stress, diet, disease processes, factors affecting mortality and life expectancy, biological responses to environmental stressors, and aging of ancestors is concerned.

Bone density achieved in early childhood is the major determinant of risk of osteoporotic fracture. Up to 60 per cent of women suffer osteoporotic fractures as a result of low bone density, which is under strong genetic control acting through effects on bone turnover (Morrison *et al.,* 1994). This genetic component can be ascribed to a simple allelic change in the receptor for 1,25 dihydroxy-vitamin D, one of the hormones controlling calcium metabolism (Mundy, 1994). Irrespective of the molecular biological or physiological mechanisms, Morrison and co-workers (1994) have identified a gene locus associated with a major part of the strong genetic effect on bone density and indeed more than half of the adjusted population variation in bone density.

A major theoretical focus in skeletal research concerns delineation and explanation of human skeletal and dental responses to changing subsistence patterns involved in the transition from hunting and gathering to an agricultural economy based primarily on corn. Traditionally, this transition has been viewed as a positive additive innovation that provided a dependable food supply, increased population density, and an improved lifestyle. In the light of recent osteological studies, however, this view requires qualification. Certain hazards accompanied over-dependence on this food source. The transition to agriculture was followed by increase in infectious diseases, pre-adult mortality, dental defects, dental caries, cranial pathology, and post-cranial periosteal reactions (Cook, 1979; Lallo *et al.,* 1977; Larson, 1981 a,b). Skeletal and dental health declined and individual size and robusticity was reduced. Given the importance of evaluating the effects of various subsistence strategies on health, studies such as these bear direct relevance to modern populations and current world problems.

Data acquired through skeletal research compliments that obtained from other archaeological sources or historical records. Consider, for example, variables, which measure patterns of wear on prehistoric dentitions. These provide information about diet and techniques of food preparation (Molnar, 1972). From an archaeological standpoint, dietary information based on the analysis of dental attrition is of considerable value since it offers an independent check against reconstructions of prehistoric subsistence based on the analysis of floral, faunal, and artifactual evidence (Walker, 1978).

Enamel hypoplasia is discernible as pits or shallow transverse grooves, especially in the canines and molars; the tooth is only slightly affected in most of these cases. This hypoplasia has been reported in *Australopithecus crassidens* –

aranthropus (Robinson, 1952), *Telanthropus capensis* (Clement, 1956), *Homo erectus – Pithecanthropus pekinensis* (Weidenreich, 1937), and *Spy* II – *Neanderthal* man (Brothwell, 1963).

Hypercementosis is found to be present in fossil men: three *Neanderthal* specimens show some root swelling, for instance, *Gibraltar* I, *La Ferrassie* I, *Monsempron*; *Rhodesian* man shows beginning of apical hypercementosis resulting from a low-grade infection of the tooth (Stones, 1954).

Dental caries has been reported in the australopithecine apes. Earliest examples of well defined caries have been found in the maxillary I and II molars of *Australopithecus crassidens* (Robinson, 1952); *Telanthropus capensis* (Clement, 1956); *Es Skhul* (Brothwell, 1961); signs of caries in *Homo erectus* (*Pithecanthropus*) (Brodrick, 1948); and eleven carious teeth in Upper Pleistocene *Rhodesian* I (Brothwell, 1963). Upper Palaeolithic teeth from Europe show some evidence of caries. Mesolithic groups reveal presence of carcinogenic factors (refined carbohydrates, honey, and pulpy plants) well before the Neolithic revolution (i.e., cereal cultivation and dietary refinement). Evidence of periodontal infection is found during the *Neanderthal* times.

Neanderthal Man

Many of the vertebrae, especially the cervical, of *La-Chapelle-aux-Saints* exhibit arthritic processes along the margins of their bodies (Straus and Cave, 1953-61): odontoid process is abnormally deflected from the median plane, articular surfaces of the atlanto-axial joints exhibit osteoarthritis change, other vertebrae are affected by a deforming osteoarthritis. As a consequence, the vertebrae are reduced in height. There also exists a flattening and deformation of the right mandibular condyle – osteoarthritis of the temporo-mandibular joint (peculiar flattening of the right condyle). Straus and Cave (1953-61) observed even more severe flattening and deformation in the *La Quina* and *La Ferrassie* mandibles, also obviously the result of temporo-mandibular osteoarthritis. Both these condyles are pathologically distorted.

Alveolar borders of the *La Chapelle* skull underwent extensive resorption during life resulting in loss of most of the teeth ante mortem – result of prolonged and extensive periodontal disease. Spinal and coaxial osteoarthritis is present in other *Neanderthal* specimens too. Arthritis is an extremely ancient malady and goes back to the Mesozoic era where it was found in the giant reptiles now extinct (Moody, 1923).

Bone Tumour

Bone tumour on the Kanam mandible may be responsible for the ascription of a *Homo sapiens* type of chin to this specimen (Montagu, 1957).

General

Considerations of the domestication process and the cultural dynamics involved are important, but somewhat less attention has been given to the dietary utilization of resources. Especially lacking is the consideration of the synergistic effects of the presence of the pathogens in animals and human susceptibility to diseases. Mixed farming and pastoral adaptations have been characteristic of some areas for significant periods of time. Since pathogens and human populations have adapted to each other through time, the epidemiological perspective should be productive in this regard. Local and regional variations in the resources, population distribution, environmental factors affecting health, nutritional adequacy of food resources, and the nature of the pathogens still must be considered.

Skeletal populations occasionally lack association with organic archaeological refuse sufficient to permit neat discriminations between stages of economic change. In a few instances, associated economies are postulated almost entirely from associated artifacts and site distributions supplemented by data from skeletons themselves.

Palaeopathological studies suffer from the facts that skeletal populations are almost always skewed representations of living groups and that the skewing may not be the same in any two populations. They suffer from the fact that different analyses have been performed in different regions and often on different populations from the same region, making both inter-regional comparisons and inter-population comparisons within a region more difficult.

Angel (1966) found the growth of Greek culture between 800 AD and 500 BC connected with such phenomenon as an increase in body size, life-span, and population volume and a decrease in arthritis, dental pathology, osteoporosis and infant mortality; also noticed a decrease in health after 400 BC. Schultz (1981) found a high degree of infestation with plasmodia filaria and trypanosome in the ancestor's of man in view of the biological relationships.

Literature Cited

Abbate, E., A. Albianelli, A. Azzaroli, M. Benvenuti, B. Tesfamariam, P. Bruni, *et al.* 1998. A one million year old *Homo* cranium from Danakil (Afar) depression of Eritrea. *Nature* 393: 458.

Ackerknecht, E.H. 1962. Palaeopathology. In *Anthropology Töday: Selections* (edited by Sol Tax). Canada: Phoenix Books. pp. 39-45.

Acsadi, G. and J. Nemeskeri 1970. *History of Human Life Span and Mortality.* Budapest: Akademiai Kaido. African populations and the evolution of human mitochondrial DNA. *Science* 253: 1503-1507.

Aguirre, E. 1975. *Kenyapithecus* and *Ramapithecus.* In *Palaeo-Anthropology–Morphology and Palaeoecology* (edited by R.H. Tuttle). The Hauge: Mouton Publishers. pp. 99-104.

Aiello, L.C. 1994. Variable but singular. *Nature* 368:399-400.

Anderson, D.M., M. Pearson, A. Fisher and D. Zieglowsky 1980. *Planning Seminar on Ancient Burial Grounds.* Special Publication, Office of State Archaeologist of Iowa, Iowa City, USA.

Andrews, P.J. 1981. Hominoid habitats of the Miocene. *Nature* 289: 749.

Andrews, P.J. 1982. Hominoid evolution. *Nature* 295:185-186.

Andrews, P.J. 1986. Fossil evidence on human origins and dispersal. Cold Spring Harbor Symposium, *Quantitative Biology* 51:419-428.

Andrews, P.J. 1992a. An ape from the south. *Nature* 356:106.

Andrews, P.J. 1992b. Evolution and environment in the Hominoidea. *Nature* 360:641-646.

Andrews, P.J. and J.E. Cronin 1982. The relationship of *Sivapithecus* and *Ramapithecus* and the evolution of the orangutan. *Nature* 297:541-546.

Andrews, P.J. and J. Tekkaya 1980. A revision of the Turkish Miocene hominoid *Sivapithecus meteai*. *Palaeontology* 23:85-95.

Angel, J.L. 1966. Porotic hyperostosis, anemias, malarias and marshes in the prehistoric eastern Mediterranean. *Science* 153:760-773.

Angel, J.L. 1969. The basis of palaeodemography. *American Journal of Physical Anthropology* 30:427-437.

Angel, J.L. 1981. History and development of paleopathology. *American Journal of Physical Anthropology* 56:509-515.

ApSimon, A.M. 1980. The last *Neanderthal* in France? *Nature* 287:271-272.

Arambourg, C. 1963. *Le Gisement De Ternifine*. Archives de l'Institut de Palaeontologie Humaine 32.

Arensburg, B. 1973. *The People in the Land of Israel from the Epi-Palaeolithic to present times.* Ph.D. Thesis, Tel Aviv University, Israel.

Armelagos, G.J., K.H. Jacobs and D.L. Martin 1981. Death and demography in prehistoric Nubia. In *Mortality and Immortality: The Anthropology and Archaeology of Death* (edited by S.C. Humphreys and H. King). New York: Academic Press. pp. 33-57.

Arsuaya, J.-L., I. Martinez, A. Garcia, J.-M. Carretero and E. Carbonell 1993. Three new human skulls from the Sima de los Huesos middle Pleistocene site in Sierra de Atapuerca, Spain. *Nature* 362:534-537.

Asfaw, B., Y. Beyenne, G. Suwa, R.C. Walter, T.D. White, G.W. Gabriel and T. Yemane 1992. The earliest Acheulean from Konso-Gardula. *Nature* 360:732-735.

Asfaw, B., T.D. White, C.O. Lovejoy, G. Suwa and S. Simpson 1999. *Australopithecus garhi*: a new species of early hominid from Ethiopia. *Science* 284: 629-635.

Aufderheide, A.C. and C. Roudriguez-Martin 1998. *The Cambridge Encyclopaedia of Human Paleopathology*. Cambridge: Cambridge University Press.

Avers, C.J. 1974. *Evolution*. New York/London: Harper & Row Publishers.

Ayala, F. 1976. *Molecular Evolution.* Massaschusetts: Sinauer.

Badam, G.L. 1979. *Pleistocene Fauna of India with Special Reference to the Siwaliks.* Poona: Deccan College Research Institute. pp 260.

Badam, G.L., R.K. Ganjoo, Salahuddin and S.N. Rajaguru 1986 Evaluation of fossil hominid—the maker of late Acheulian tool at Hathnora, Madhya Pradesh, India. *Current Science* 55: 143-145.

Barker, B. 1973. An additional aid to sexing aboriginal skulls b measuring mandibular molars at the cemento-enamel junctior *Archaeology and Physical Anthropology of Oceania* 8: 125-133.

Bartstra, G.-J, S.G. Keates, Basoeki and B. Kallupa 1991. On the dispersion of *Homo sapiens* in Eastern Indonesia: The Palaeolithic of South Sulawesi. *Current Anthropology* 32:317-321.

Begun, D.R. 1994. Relations among the great apes and humans: new interpretations based on fossil great ape, *Dryopithecus*. *Yearbook of Physical Anthropology* 37: 11-63.

Beyene, Y., G. Suwa and B. Asfaw 1992. The earliest Acheulian from Konso-Gardula. *Nature* 360: 732-735.

Bilsborough, A. 1971. Evolutionary change in the hominid maxilla. *Man* 6: 473-485.

Bilsborough, A. 1979. Reply to "Implications of border cave skeletal remains for later Pleistocene human evolution". *Current Anthropology* 20:27.

Bilsborough, A. 1992. *Human Evolution*. London: Chapman and Hall.

Bodmer, W.F. 1983. Gene clusters and genome evolution. In *Evolution from Molecules to Men* (edited by D.S. Bendall). Cambridge: Cambridge University Press. pp. 197-207.

Borah, W. 1964. America as model: The demographic impact of European expansion upon the non-European world. 35th Congress Internacional de Americanistas, Mexico, 1962. Actas y Memorias, Mexico 3:381-382.

Boule, M. and H.V. Vallois 1957. *Fossil Men: A Textbook of Human Paleaontology*. London: Thames and Hudson.

Bowcock, A.M., J.R. Kidd, J.L. Mountain, J.M. Herbert, L. Carotenuto *et al* 1991. Drift, admixture, and selection in human evolution: a study with DNA polymorphisms. *Proc. National Academy of Sciences,* U.S.A. 88:839-843.

Brace, C.L. 1964. The fate of the 'Classic' *Neanderthals*: a consideration of hominid catastrophism. *Current Anthropology,* 5:3-43.

Braüer, G. 1984. The 'Afro-European sapiens-hypothesis', and hominid evolution in East Asia during the late middle and upper Pleistocene. Cour. Forsch. Senckenberg 69:145-165.

Braüer, G. 1989. The evolution of modern humans: a comparison of the African and non-African evidence. In *The Human Revolution: Behavioural and Biological Perspectives on the Origins of Modern Humans* (edited by P. Mellars and C. Stringer). Edinburgh: Edinburgh University Press.

Brodrick, A.H. 1948. *Early Man*. London: Hutchinson.

Brooks, A.S. 1992. Modern human origins—What's new with what's old. *Anthro. Notes* 14(3).

Broom, R. and J.T. Robinson 1950a. Man contemporaneous with Swartkrans ape-man. *American Journal of Physical Anthropology* 8:151-155.

Broom, R. and J.T. Robinson 1950b. Sterkfontein ape-man, *Plesianthropus*. Transvaal Museum Memoir 4.

Brose, D.S. and M.H. Wolpoff 1971. Early upper palaeolithic man and late middle palaeolithic tools. *American Anthropologist* 73:1156-1194.

Brothwell, D.R. 1961a. The palaeopathology of early British man: an essay on the problems of diagnosis and analysis. *Journal of the Royal Anthropological Institute* 91:318.

Brothwell, D.R. 1961b. The people of Mount Carmel: A reconsideration of their position in human evolution. Proceedings of Prehistory Society 27: 155.

Brothwell, D.R. 1963a. *Digging up bones*. London: British Museum of Natural History.

Brothwell, D.R. 1963b. The palaeopathology of Pleistocene and more recent mammals. In *Science in Archaeology* (edited by D.R. Brothwell and E. Higgs). London: Thames and Hudson.

Brothwell, D.R. 1963c. The macroscopic dental pathology of some earlier human populations. In *Dental Anthropology* (edited by D.R. Brothwell). London: Pergamon Press. pp. 271-288.

Brothwell, D.R. 1967. The evidence for neoplasms. In *Diseases in Antiquity* (edited by D.R. Brothwell and A.T. Sandison). Springfield: Thomas.

Brown, B., W.K. Gregory and M. Hellman 1924. On three incomplete anthropoid jaws from the Siwaliks, India. American Museum Novitates, No. 130: 1-9.

Brown, B. and S.C. Ward 1988. Facial and basicranial relationships in *Sivapithecus* and *Pongo*. In *Biology of the Orangutan* (edited by J.H. Schwartz). Oxford: Oxford University Press. pp. 247-260.

Brown, W.M. 1980. Polymorphism in mitochondrial DNA of humans as revealed by restriction endonuclease analysis. *Proceedings National Academy of Sciences,* USA 77:3605-3609.

Brunet, M. 1996. New species of early human reported found in desert area of Chad. *The Providence Journal-Bulletin,* Thursday, May 23. pp. A 13.

Buikstra, J.E. 1977. Differential diagnosis: an epidemiological model. *Yearbook of Physical Anthropology* 20: 316-328.

Buikstra, J.E. and D.C. Cook 1980 Palaeopathology: an American account. *Annual Review of Anthropology* 9:433-470.

Buikstra, J.E. and L.W. Königsberg 1985. Palaeodemography: critiques and controversies. American Anthropologist 87:316-333.

Buikstra, J.E. and J.H. Mielke 1985. Demography, diet and health. In *The Analysis of Prehistoric Diets* (edited by R.I. Gilbert, Jr. and J.H. Mielke). New York: Academic Press Inc. pp. 359-422.

Burenhult, G. 1993. *The First Humans: human origins and history to 10,000 B.C.* (American Museum of Natural History – The Illustrated History of Mankind). San Francisco: Harper.

Campbell, B. 1962. The systematics of man. *Nature* 194:225-232.

Campbell, B. 1964. Just another "man-ape"? *Discover* 25:37-38.

Campbell, B. 1965. The nomenclature of the Hominidae including a definitive list of hominid taxa. Royal Anthropological Institute of Great Britain and Ireland. Occasional Paper No. 22.

Campbell, B.G. 1967. *Human Evolution.* Cicago, Illinois: Aldine Publishing Company.

Campbell, B.G. and J.D. Loy 1996. *Humankind Emerging.* 7th edition. New York: Harper Collins College Publishers.

Cann, R.L 1988. DNA and human origins. *Annual Review of Anthropology* 17:127-143.

Cann, R.L., M. Stoneking and A.C. Wilson 1987. Mitochondrial DNA and human evolution. *Nature* 325:31-36.

Cartmill, M., D. Pilbeam and G. Isaac 1986. One hundred years of palaeoanthropology. *American Scientist* 74:410-420.

Cassidy, C.M. 1984. Skeletal evidence for prehistoric subsistence adaptation in the Central Ohio River Valley. In *Palacopathology at the Origins of Agriculture* (edited by M.N. Cohen and G.J. Armelagos). Orlando: Academic Press, Inc. pp. 307-345.

Cavalli-Sforza, L.L. and A.W.F. Edwards 1967 (1971). Phylogenetic analysis: models and estimation procedures. *American Journal of Human Genetics* 19: 233-257.

Cavalli-Sforza, L.L., P. Menozzi and A. Piazza 1993. Demic expansions and human evolution. *Science* 259:639-646.

Cavalli-Sforza, L.L. and A. Piazza 1975. Analysis of evolution: evolutionary rates, independence and treeness. *Theoretical Population Biology* 8:127-165.

Chaimanee, V., V. Suteethorn, J.-J. Jaeger and S. Ducrocq 1997. A new late Eocene anthropoid primate from Thailand. *Nature* 385:429-431.

Chang, K. 1962. New evidence on fossil man in China. *Science* 136: 749–760.

Chase, P.G. and H.L. Dibble 1990. On the emergence of modern humans. *Current Anthropology* 31: 58 -59.

Chopra, S.R.K. 1968a. The early man in Sivaliks I. *Akashvani* 33(45): 5.

Chopra, S.R.K. 1968b. The early man in Sivaliks II. *Akashvani* 33(47): 9.

Chopra, S.R.K. 1974. *Palaeoprimatological Studies in India with Special Reference to Recent Finds in the Sivaliks.* Presidential Address, Section of Anthropology and Archaeology, 61st Session of the Indian Science Congress, Nagpur, Pt. II, pp. 1-16.

Chopra, S.R.K. 1976. Primatological studies in India. In *Anthropology in India*, edited by H.K. Rakshit. Calcutta: Anthropological Survey of India. pp. 124-143.

Chopra, S.R.K. 1978. New fossil evidence on the evolution of *Hominoidea* in the Sivaliks and its bearing on the problem of the evolution of early man in India. *Journal of Human Evolution* 7: 3-9.

Chopra, S.R.K. 1979a. *Early Man in North West India*. New Delhi: Allied Publishers Private Limited.

Chopra, S.R.K. 1979b. Palaeontological evidence bearing on the problem of human origins in North-West India. In *Early Man in North West India* (edited by S.R.K. Chopra). New Delhi: Allied Publishers Private Limited. pp. 1-19.

Chopra, S.R.K. 1983. Significance of recent hominoid discoveries from the Siwalik Hills of India. In *New Interpretations of Ape and Human Ancestry* (edited by R.L. Ciochon and R.S. Corruccini). New York: Plenum Publishing Press. pp. 539-557.

Chopra, S.R.K. 1984. *Evolution of Early Man in India—Perspectives and Issues.* Second M.R. Sahni Memorial Lecture, Palaeontological Society of India, Lucknow March 1, 1984. pp. 1-15.

Chopra, S.R.K. 1987. *Fossil Apes and Men in India—Status and Prospects*. Dr. Panchanan Mitra Memorial Lecture, The Asiatic Society, Calcutta.

Chopra, S.R.K. and S. Kaul 1975. New fossil *Dryopithecus* material from the Nagri beds at Haritalyangar (H.P.), India. In *Contemporary Primatology* (edited by S. Kondo, M. Kawai and A. Ehara). Basel and New York: S. Karger. pp. 2-11.

Chopra, S.R.K. and S. Kaul 1979. A new species of *Pliopithecus* from the Indian Sivaliks. *Journal of Human Evolution* 8: 475-477.

Chopra, S.R.K., S. Kaul and R.N. Vasishat 1979. Miocene tree shrews from the Indian Sivaliks. *Nature* (Lond.) 281: 213-214.

Chopra, S.R.K. and R.N. Vasishat 1979. Sivalik fossil tree shrew from Haritalyangar, India. *Nature* 281 (No. 5728): 214–215.

Chopra, S.R.K. and R.N. Vasishat 1980a. A new Mio-Pliocene *Indraloris* (Primate) material with comments on the taxonomic status of *Sivanasua* (Carnivore) from the Sivaliks of the Indian subcontinent. *Journal of Human Evolution* 9: 129-132.

Chopra, S.R.K. and R.N. Vasishat 1980b. Première indication de la prèsence dans le Mio-Pliocène des Siwaliks de l'Inde d'un Primate adapidè, *Indoadapis shivaii*, nov. sp. C.R. Acad. Sc. Paris, Sèrie D 290: 511-513.

Chopra, S.R.K. and R.N. Vasishat 1984. Fossil adapids from Kangra district of Himachal Pradesh, India. Xth Congress International Primatological Society of India, Nairobi, Kenya. Abstract.

Ciochon, R. 1995. The earliest Asians yet. *Natural History* 104(12): 50-54.

Ciochon, R.L. and A.B. Chiarelli 1980. Palaeobiogeographic perspectives on the origin of the platyrrhini. In *Evolutionary Biology of the New World Monkeys and Continental Drift* (edited by R.L. Ciochon and A.B. Chiarelli). New York: Plenum Publishing Corporation. pp. 459-494.

Ciochon, R.L. and J.G. Fleagle 1993. *The Human Evolution*—Source Book. Englewood Cliffs, New Jersey: Prentice Hall, Inc.

Clark, J.D. 1958. Early man in Africa. *Scientific American,* July issue.

Clark, W.E. Le Gros 1934. *Early Forerunners of Man*. London.

Clark, W.E. Le Gros 1947. Observations of the anatomy of the fossil australopithecinae. *Journal of Anatomy* 81:1-35.

Clark, W.E. Le Gros 1959. *The Antecedents of Man*. Edinburgh University Press.

Clark, W.E. Le Gros 1965. *History of Primates*. London: British Museum of Natural History.

Clark, W.E. Le Gros and L.S.B. Leakey 1951. *The Miocene Hominoidea of East Africa*. Brit. Mus. Nat. Hist. London: Fossil Mammals of Africa No. 1. pp. 1-117.

Clark, W.E. Le Gros and D.P. Thomas 1951. Associated jaws and limb bones of *Limnopithecus macinnesi*. British Museum of Natural History, London: Fossil Mammals of Africa. 3:1-27.

Clarke, R. 1997. Fantastic fossil foot found. *Anthro Bulletin,* American Museum of Natural History.

Clarke, R.J. 1985. In *Ancestors: The Hard Evidence*. (edited by E. Delson). New York: Alan R. Liss. pp. 301-305.

Clement, A.J. 1956. Caries in the South African ape-man: some examples of undoubted pathological authenticity believed to be 800,000 years old. *British Dental Journal* 102: 4

Colbert, E.H. 1935. Siwalik mammals in American Museum of Natural History. Transactions of the American Philosophical Society, n.s., 26: 1-401.

Conroy, G.C. 1972. Problems in the interpretation of *Ramapithecus*: with special reference to anterior tooth reduction. *American Journal of Physical Anthropology* 37: 41-46.

Conroy, G.C., M. Pickford, B. Senut, J. van Couvering and P. Mein 1992. *Otavipithecus namibiensis*, first Miocene hominoid from southern Africa. *Nature* 356:144-148.

Conroy, G.C. and D.R. Pilbeam 1975. *Ramapithecus*: a review of its hominid status. In *Palaeoanthropology-Morphology and Palaeoecology* (edited by R.H. Tuttle). The Hauge: Mouton Publishers. pp. 59-86.

Cook, D.C. 1979. Subsistence base and health in prehistoric Illinois Valley: evidence from the human skeleton. *Medical Anthropology* 3:109-124.

Cooke, H.B.S. 1963. Pleistocene mammal faunas of Africa with particular reference to southern Africa. In *African Ecology and Human Evolution* (edited by F. Howell and Bourliere). Chicago: Aldine-Atherton, Inc. pp. 65-116.

Coon, C.S. 1963. *The Origin of Races*. New York: Alfred A. Knopf.

Coppa, A. 1988. Mortality trends at Geili (Central Sudan) from Meroitic to Christian periods. *El Geili: The History of a Middle Nile Environment 7000 B.C.-A.D. 1500* (edited by I. Caneva). Oxford: B.A.R. Cambridge Monographs in African Archaeology 29:227-273.

Coppens, Y. Hominization in general. In *Methodology and African Prehistory*. pp. 200-419.

Coppens, Y. 1961. Découverte d'un Australopitheciné dans le Villafranchien du Tchad. Comptes Rend. *Seances Acad. Sci.*, (Paris) 252:3851-3852.

Coppens, Y. 1983. Systematique, phylogenie, environment et culture des australopitheques, hypotheses et synthese. *Bull. Mém. Soc. Anthropol.* (Paris), Ser. XIII, 10:273-284.

Cornwall, I.W. 1956. *Bones for the Archaeologist*. London: Phoenix.

Crenson, M. 1996. Anthropologists come face to face with a possible human ancestor. *The News and Observer,* Thursday July 25.

Culotta, E. 1995. Asian hominids grow older. *Science* 270: 1116-1117.

Dart, R.A. 1925. *Australopithecus africanus*: the man ape of South Africa. *Nature* 115:195-199.

Dart, R.A. 1960. The status of *Gigantopithecus*. *Anthropologisches Anzeiger* 24: 139-145

Darwin, C. 1871. *The Descent of Man*. London: John Murray.

Day, M.H. 1977. *Guide to Fossil Man*. 3rd ed. London: Cassell & Company Limited.

Day, M.H. 1986. *Guide to Fossil Man*. 4th ed. London: Cassell & Company Limited.

de Chardin, P.T. 1952. The idea of fossil man. In *Anthropology Today* (edited by A.L. Kroeber). Chicago, Illinois.

de Grouchy, J. C. Turleau and C. Finaz 1978. Chromosomal phylogeny of the primates. *Annual Review of Genetics* 12:289-328.

de Lumley, H. and A. Sonakia 1985a. Contexte stratigraphique et archaeologique de l'homme de la Narmada, Hathnora, Madhya Pradesh, Inde. *L'Anthropologie (Paris)* 89:3-12.

de Lumley, M.-A. and A. Sonakia 1985b. Premiere decouverte d'un *Homo erectus* sur le continent Indien a Hathnora, dans la Moyenne Vallee de la Narmada. *L'Anthropologie* (Paris) 89:13-61.

de Terra, H. and T.T. Patterson 1939. *Studies on the Ice Age in India and Associated Human Cultures*. Carnegie Institute Washington Publication No. 493: 1-354.

de Villatta, J.F. and M. Crusafont Pairo 1944. Dos nuevos Antropomorfos del Mioceno espânol, y su situación de la moderna sistematica de los Simidos. Notes y Commun. Inst. Geol. Min. Espana 13:7-25.

de Vore, J. 1992. *Molecular Applications in Biological Anthropology*. Cambridge: University Press.

Dickel, D.N., P.D. Schulz and H.M. McHenry 1984. Central California: prehistoric subsistence changes and health. In *Palaeoanthropology at the Origins of Agriculture* (edited by M.N. Cohen and G.J. Armelagos). Orlando: Academic Press Inc. pp 439-461.

Dobyns, H.F. 1966. Estimating aboriginal American population: an appraisal of techniques with a new hemispheric estimate. *Current Anthropology* 7: 395- 416.

Dobzhansky, T. 1962. *Mankind Evolving: The Evolution of the Human Species*. New Haven: Yale University Press.

Durant, J.R. 1989. *Human Origin*. Oxford: Clarendon Press.

Dutrillaux, B. 1976. Study of human X chromosomes with the 5-BrdU-acridine orange technique. Application to X chromosome pathology. *Chromosomes Today* 5:395-407.

Dutrillaux, B. 1979. Chromosomal evolution in Primates: Tentative phylogeny from *Microcebus murinus* (Prosimian) to Man. *Human Genetics* 48: 251-314.

Dutrillaux, B. and J. Couturier 1983. The ancestral karyotype of Carnivora: comparison with that of platyrrhine monkeys. *Cytogenetics and Cell Genetics* 35:200-208.

Dutrillaux, B., E. Viegas-Pequignot, E. Dubos and R. Masse 1978. Complete or almost complete analogy of chromosome banding between baboon (*Papio papio*) and Man. *Human Genetics* 43:37-46.

Dutrillaux, B., E. Viegas-Pequignot, J. Couturier and G. Chauvier 1978. Identity of euchromatic bands from man to Cercopithecidae (*Cercopithecidae aethiops, Cercopithecus sabaeus, Erythrocebus patas, Miopithecus talapoin*). *Human Genetics.*

Dutta, A.K., P.K. Basu and M.V.A. Sastry 1976. On the new finds of hominids and additional finds of pongids from the Siwaliks of Ramnagar area, Udhampur District, J & K State. *Indian Journal of Earth Sciences* 3: 234-235.

Eckhardt, R.B. 1988. On early hominid adaptation to heat stress. *Current Anthropology* 29: 493.

Eckhardt, R.B. 1975. *Gigantopithecus* as a hominid. In *Palaeoanthropology-Morphology and Palaeoecology* (edited by R.H. Tuttle). The Hauge: Mouton Publishers. pp. 105-129.

Egozcue, J. 1969. Primates. In *Comparative Mammalian Cytogenetics* (edited by K. Benirschke) New York: Springer-Verlag. pp. 357-389.

Etler, D.A. 1984. The fossil hominids of Lufeng, Yunnan Province, The People's Republic of China: a series of translations. *Yearbook of Physical Anthropology* 27: 1-55.

Ewer, R.F. 1957 Faunal evidence on the dating of the Australopithecinae. Procs. III Pan-African Cong. Prehistory, northern Rhodesia, 1955. London: Chatto and Windus. pp. 135-142.

Ewer, R.F. 1973. *The Carnivores*. Ithaca: Cornell University Press.

Ewer, R.F. and H.B.S. Cooke 1964. The Pleistocene mammals of southern Africa. In *Ecological Studies in Southern Africa* (edited by D.H.S. Davis). *Monographiae Biologicae* 14:35-48.

Fairservis, W.A. 1975. *The Roots of Ancient India: The Archaeology of Early Indian Civilisation.* 2nd ed. Chicago: University of Chicago Press.

Falconer, H. and P.T. Cautley 1836. *Sivatherium giganteum:* a new fossil ruminant genus from the valley of the Markanda in the Siwalik branch of the sub-Himalayan mountains. *Asiatic Researches* 19: 1-24.

Falconer, H. and P.T. Cautley 1868. *Palaeontological Memoirs* (edited by R.I. Murchison). I and II.

Ferembach, D., I. Schwidetzky and M. Stloukal 1977-79. Raccomandazioni per la determinazione dell'eta e del sesso sullo scheletro. *Revista di Antropologia* 6: 5-51.

Ferishta 1360. quoted from Falconer and Cautley (1836).

Ferris, S.D. *et al.* 1981. Extensive polymorphism in the mitochondrial DNA of apes. *Procs. National Academy of Sciences,* U.S.A. 78:6319-6323.

Fleagle, J.G. 1988. *Primate Adaptation and Evolution.* New York: Academic Press.

Flynn, L.J., D.R. Pilbeam, L.L. Jacobs, J.C. Barry, A.K. Behrensmeyer and J.W. Kappelman 1990. The Siwaliks of Pakistan: time and faunas in a Miocene terrestrial setting. *Journal of Geology* 98: 589-604.

Foley, R. 1984. Early man and the Red Queen: tropical African community evolution and hominid adaptation. Prehistoric human adaptation in biological perspective. In *Hominid Evolution Community Ecology* (edited by R. Foley). London: Academic Press. pp. 85-110.

Foran, D.R., J.E. Hixson and W.M. Brown 1988. Comparisons of ape and human sequences that regulated mitochondrial DNA transcription and D-loop synthesis. *Nucleic Acids Res.* 16:5841-5861.

Franciscus, R.G. and E. Trinkaus 1988. Nasal morphology and the emergence of *Homo erectus. American Journal of Physical Anthropology* 75:517-527.

Gadamer, H.-G. and P. Vogler 1972. *Neus Anthropologie-Biologische Anthropologie* 2nd Teil. Stuttgart: Georg Thieme Verlag. dtv wissenschaft.

Gamble, C. 1994a. Human evolution: the last one million years. In *Companion Encyclopaedia of Anthropology* (edited by T. Ingold). London and New York: Routledge. pp. 79-107.

Gamble, C.S. 1994b. *Timewalkers. The Prehistory of Global Colonization.* Cambridge: Harvard University Press.

Garn, S.M. 1967. Sex difference in tooth shape. *Journal of Dental Research* 46: 1470.

Garn, S.M., P.E. Cole and W.L. van Alstine 1979. Sex discriminatory effectiveness using combinations of root lengths and crown diameters. *American Journal of Physical Anthropology* 50:115-118.

Garver, F.A. and D.W. Ftalmage 1975. Comparative immunochemical studies of primate hemoglobins. *Biochemical Genetics* 13:743-757.

Gates, R. 1948. *Human Ancestry*. Cambridge: Cambridge University Press.

Gaudry, A. 1878. *Les enchainements du monde animal. Mammiferes tertiaires,* Paris. pp. 236.

Gaudry, A. 1890. *Le Dryopitheque. Mém. de la Soc. géolg, de France, Paleontologie,* Paris No. 1.

Gaur, R. and R.N. Vasishat 1989. Hominoid Palaeoecology of the Indian Siwaliks. In *Perspectives in Human Evolution* (S.R.K. Chopra Festschrift Volume) (edited by A. Sahni and R. Gaur). Delhi: Renaissance Publishing House. pp. 195-222.

Gaur, R., R.N. Vasishat and S.R.K. Chopra 1978. Upper Sivalik vertebrate communities of the Indian subcontinent. *Recent Researches in Geology* 5: 206-218.

Gebo, D. 2000. Cited from Gugliotta (2000).

Gervais, P. 1872. Sur un singe fossile d'espéce non encore décrite, qui a été de couvert au Monte - Bamboli (Italie). *C.R. Acad. Sci.* Paris 74:1217-1223.

van Gerven, D.P. and G.J. Armelagos 1983. "Farewell to palaeodemography?" Rumors of its death have been greatly exaggerated. *Journal of Human Evolution* 12: 353-360.

Ghosh, A.K. 1975. *Man Evolving in India*. Presidential Address, Section of Anthropology and Archaeology. *Procs. 62nd Indian Science Congress Association.* pp. 1-27.

Gibbons, A. 1996. *Homo erectus* in Java: A 250,000-year anachronism. *Science* 274:1841-1842.

Gingerich, P.D. and A. Sahni 1979. *Indraloris* and *Sivaladapis*: Miocene adapid primates from the Siwalik of India and Pakistan. *Nature* (Lond.) 279: 415-416.

Gingerich, P.D. and A. Sahni 1984. Dentition of *Sivaladapis nagri* (Adapidae) from the Late Miocene of India. *International Journal of Primatology* 5: 63-79.

Ginsburg, L. and P. Mein 1980. *Crouzelia rhodanica*, nouvelle espece de primate catarhinien, et essai sur la position systematique des pliopithecidae. *Bulletin Museum of Natural History Nat.* Paris, Ser. 4, 2(c): 57-85.

Godinot, M. and M. Mahboubi 1992. Earliest known simian primate found in Algeria. *Nature* 357:324-326.

Goldstein, M.S. 1963. The palaeopathology of human skeletal remains. In *Science in Archaeology* (edited by D.R. Brothwell and E. Higgs). London: Thames and Hudson.

Goodman, M. *et al.* May 1987. *Procs. National Academy of Sciences,* U.S.A. 84: 3307-3311.

Goodman, M. 1962. Evolution of the immunologic species specificity of human serum proteins. *Human Biology* 34:104-120.

Goodman, M. 1963. Serological analysis of the systematics of recent hominoids. *Human Biology* 35:377-390.

Goodman, M. 1964. Man's place in the phylogeny of the primates as reflected in serum proteins. In *Classification and Human Evolution* (edited by S.L. Washnurn). London: Methuen & Co. pp. 204-234.

Goodman, M. 1965. The specificity of proteins and the process of primate evolution. In *Protides of the Biological Fluids* (edited by H. Peters). Amsterdam: Elsevier Publications. pp. 70-86.

Goodman, M. 1967. Deciphering primate phylogeny from macromolecular specificities. *American Journal of Physical Anthropology* 26: 255-276.

Goodman, M. 1968. Phylogeny and taxonomy of the catarrhine primates from immunodiffusion data. A review of major findings. In *Taxonomy and Phylogeny of Old World Primates with References to the Origin of Man* (edited by B. Chiarelli). Turin: Rosenberg and Sellier. pp. 95-107.

Goodman, M. 1975. Protein sequence and immunological specificity. In *Phylogeny of Primates* (edited by F. Luckett and F. Szalay). New York: Plenum Press.

Goodman, M. 1977. Molecular evolution and natural selection in the descent to man. *Primate News* 15: 3-7.

Goodman, M. 1982. *Macromolecular Sequences in Systematic and Evolutionary Biology*. New York: Plenum Press.

Goodman, M., B.F. Koop, J. Czelusniak, D.H.A. Fitch, D.A. Tagel and J.L. Slightom 1989. Molecular phylogeny of the family of apes and humans. *Genome* 31:316-335.

Goodman, M., R.E. Tashian and J. Tashian 1976. *Molecular*

Anthropology: Genes and Proteins in the Evolutionary Ascent of the Primates. New York: Plenum Press.

Gould, S.J. 1982. The meaning of punctuated equilibria and its role in validating a hierarchical approach to macroevolution. In *Perspectives in Evolution.* (edited by R. Milkman). Cambridge: Cambridge University Press. pp. 71-93.

Greenfield, L. 1979. On the adaptive pattern of *Ramapithecus*. *American Journal of Physical Anthropology* 50:527-548.

Greenfield, L.O. 1980. A late divergence hypothesis. *American Journal of Physical Anthropology* 52: 351-365.

Gregg, J.B., J.P. Steele, L. Zimmerman, H. Ferwerda and P.S. Gregg 1981a. Otolaryngic osteopathology in 14th century mid-America. *Ann. Otology, Rhinology and Laryngology* 90:288-293.

Gregory, W.K. 1910. The Orders of mammals. *Bulletin of the American Museum of Natural History* 27:1.

Gregory, W.K. 1916. Studies on the evolution of the primates. *Bulletin of the American Museum of Natural History* 35:239.

Gregory, W.K. 1922. *The Origin and Evolution of the Human Dentition.* Baltimore: Williams and Wilkins.

Gregory, W.K. 1939a. Fossil man-apes of South Africa. *Nature* 143:25-27.

Gregory, W.K. 1939b. South African fossil man-apes and the origin of human dentition. *Journal of American Dental Association* 26:558-564.

Gregory, W.K. 1951. *Evolution Emerging: a survey of changing patterns from primeval life to man.* Vol. II. New York: The Macmillan Company.

Gregory, W.K. and M. Hellman 1926. The dentition of *Dryopithecus* and the origin of man. *Anthropological Papers of the American Museum of Natural History* 28:1.

Gregory, W.K., M. Hellman and G.E. Lewis 1938. *Fossil Anthropoids of the Yale Cambridge India Expedition of 1935.* Carnegie Institute Washington Publication No. 495: 1-27.

Gugliotta, D. 2000. Tiny fossils, big find: small primates in China give clues to evolution. *Washington Post,* March 16, 2000/San Francisco Chronicle.

Gupta, S.S., B.C. Verma and A.P. Tiwari 1982. New fossil hominoid material from the Siwaliks of Kangra District, Himachal Pradesh. *J. Palaeontological Society of India* 27: 111-115.

Gupta, V.J. 1969. Fossil primate from the lower Siwaliks of Kangra.

Research Bulletin of Panjab University, Chandigarh 20: 577-578.

Gupta, V.J., I.J. Suneja and S.R.K. Chopra 1979. Stratigraphy and palaeoecology of Quaternary sediments of North-West Himalaya, India. In *Early Man in North West India* (edited by S.R.K. Chopra). New Delhi: Allied Publishers Private Limited. pp. 66-97.

Haeckel, E. 1866. *Generelle Morphologie Der Organismen.* Berlin: G. Reimer.

Hay, R.L. and M.D. Leakey 1982. The fossil footprints of Laetoli. *Scientific American* 246:50-57.

Heberer, G. 1956. Die Fossilgeschichte der Hominoidea. In *Primatologia* (edited by H. Hofer, A.H. Schultz and D. Starck). Basel: S. Karger. Vol. 1, pp. 379-560.

Hill, A., S. Ward, A. Deino, G. Curtis and P. Drake 1992. Earliest *Homo. Nature* 355:719-722.

Hoffman, J.M. 1979. Age estimations from diaphyseal lengths: two months to twelve months. *Journal of Forensic Science* 24: 461-469.

Hofman, M.A. 1983. Encephalization in hominids: evidence for the model of punctuationalism. *Brain, Behaviour and Evolution* 22: 102-117.

Hofmann, A. 1893. Die Fauna von Göriach. Abh. d. K.K. geol. Reichsanstalt, 1. Wien.

Hooton, E.A. 1954. *Up from the Ape.* New York: The Macmillan Company.

Hopwood, A.T. 1933 Miocene primates from Kenya. *Journal of Linnean Society* (Zool) 38:437-464.

Howell, F.C. 1964. The hominization process. In *Horizons of Anthropology* (edited by Sol Tax). Chicago: Aldine Publishing Company. pp. 140-146.

Howell, F.C. 1973. *Early Man.* New York: Time-Life Books.

Howell, F.C. 1978. Hominidae. In *Evolution of African Mammals* (edited by V.J. Maglio and H.B.S. Cooke). Cambridge: Harvard University Press. pp. 154-248.

Howell, F.C. 1982a. Origins and evolution of African Hominidae. In *The Cambridge History of Africa* Vol. 1 (edited by J.D. Clarke). Cambridge: Cambridge University Press. pp. 70-156.

Howell, F.C. 1982b. Pers. Comm. to Nice UISPP Congress delegates.

Howell, N. 1976. Toward a uniformitarian theory of human

palaeodemography. In *The Demographic Evolution of Huaman Population* (edited by R.H. Ward and K.M. Weiss). New York: Academic Press Inc. pp. 25-40.

Howells, W.W. 1947. *Mankind So Far*. London: Sigma Books Limited (The American Museum of Natural History, Science Series).

Howells, W.W. 1960. The distribution of man. Scientific American September. pp 113-127. Also in *Human Variation and Origins: an introduction to human biology and evolution.* (Readings from Scientific American). San Francisco/London: W.H. Freeman and Co. pp. 215-223.

Howells, W.W. 1962. *Ideas on Human Evolution: Selected Essays 1949-1961*. Cambridge, Mass: Harvard University Press.

Howells, W.W. 1967. *Mankind in the Making*, revised edition. Doubleday, Garden City.

Howells, W.W. 1974. *Neanderthals*: names, hypothesis and scientific method. *American Anthropologist* 76:24-38.

Howells, W.W. 1976. *Homo erectus.* In *Avenues to Antiquity—Readings from Scientific American.* San Francisco: W.H. Freeman and Company. pp. 28-35.

Howells, W.W. 1978. Position phylétique de l'homme de *Neanderthal*. In *Les Origines Humaines et les Epoques de l'Intelligence*. Paris: Masson. pp. 217-235.

Hrdlicka, A. 1935. The Yale fossils of anthropoid apes. *American Journal of Science* series 5, 29: 34.

Hrdlicka, A. 1980. *The Skeletal Remains of Early Man.* Washington, D.C.: Smithsonian Miscellaneous Collections No. 83.

Huang, W., R. Ciochon, Y. Gu, R. Larick, Q. Fang, H. Schwarcz, C. Yonge, J. de Vos and W. Rink 1995. Early *Homo* and associated artefacts from Asia. *Nature* 378: 275-278.

Huang, W., Y. Gu, R. Ciochon, R. Larick, Q. Fang, J. de Vos, H. Schwarcz, W. Rink and C. Yonge 1996. Whose teeth? Reply to Schwarcz and Tattersall (1996). *Nature.*

Hublin, J.-J. 1982. Les Anténéandertaliens: Présapiens ou Prénéandertaliens. *Geobios mem.* spec. 6:345-357.

Hürzeler, J. 1954. Contribution á l'odontologie et á la phylogénése du genee *Pliopithecus* Gervais. *Ann. Pal.* 40:1-63.

Hürzeler, J. 1958. *Oreopithecus bambolii* Gervais: a preliminary report. *Verh. Naturforsch. Ges.* Basel 69:1-48.

Hürzeler, J. 1968. Questions et reflexions sur l'histoire des anthropomorphes. *Ann. Pal.* 54:195-233.

Huxley, T. 1863. *Evidence as to Man's Place in Nature.* London.

Ingold, T. 1994. *Companion Encyclopaedia of Anthropology.* London and New York: Routledge.

Ingram, V.M. 1961. Gene evolution and the hemoglobins. *Nature* (Lond.) 189:704.

Jacob, T. 1985. Morphology and palaeoecology of early man in Java. In *Palaeoanthropology, Morphology and Palaeoanthropology* (edited by R.H. Tuttle). The Hague: Mouton. pp. 311-325.

Johanson, D.C. 1980. Early African hominid phylogenesis: a re-evaluation. In *Current Arguments on Early Man* (edited by L.-K. Königsson). Oxford: Pergamon Press. pp. 31-69.

Jolly, C.J. 1970. The seed eaters: a new model of hominid differentiation based on a baboon analogy. *Man* 5:5-26.

Jones, K.W., J. Prosser, G. Corneo, E. Ginelli and M. Bobrow 1972. Satellite DNA, constitutive heterochromatin and human evolution. In *Modern Aspects of Cytogenetics: Constitutive Heterochromatin in Man* (edited by F.K. Pfeiffer). Stuttgart: Schattauer Verlag. *Symposia Medica Hoechst* 6: 45-61.

Jones, S., R. Martin and D. Pilbeam 1992. *The Cambridge Encyclopaedia of Human Evolution.* Cambridge: Cambridge University Press.

Kahn, P. and A. Gibbons 1997. DNA from an extinct human. *Science* 287: 176-178.

Kappelman, J., J. Kelley, D. Pilbeam, K.A. Sheikh, S. Ward, M. Anwar, J.C. Barry, B. Brown, P. Hake, N.M. Johnson, S.M. Raza and S.M.I. Shah 1991. The earliest occurrence of *Sivapithecus* from the middle Miocene Chinji Formation of Pakistan. *Journal of Human Evolution* 21: 61-73.

Kaul, S. and R.N. Vasishat 1981. On the status of *Sivapithecus indicus* material from the lower Siwaliks of Kangra District, H.P. *Current Science* 50: 534-535.

Kay, R. 1983. *Ramapithecus* reclaimed. *The Sciences* Jan/Feb:26-27.

Kay, R.F. 1982a. *Sivapithecus simonsi*, a new species of Miocene hominoid, with comments on the phylogenetic status of the ramapithecinae. *International Journal of Primatology* 3: 113-173.

Kay, R.F. 1982b. Sexual dimorphism in ramapithecinae. *Procs. National Academy of Sciences,* U.S.A. 79: 209-212.

Kay, R.F. and E.L. Simons 1983. A reassessment of the relationship between later Miocene and subsequent hominoidea. In *New Interpretations of Ape and Human Ancestory* (edited by R.L.

Ciochon and R.S. Corruccini). New York: Plenum Publishing Corporation. pp. 577-624.

Kelley, J. 1986. Species recognition and sexual dimorphism in *Proconsul* and *Rangwapithecus*. *Journal of Human Evolution* 15: 461-495.

Kelley, J. 1993. Taxonomic implications of sexual dimorphism in *Lufengpithecus*. In *Species, Species Concepts, and Primate Evolution* (edited by W.H. Kimbel and L.B. Martin). New York: Plenum Press. pp. 429-458.

Kelley, J. and D. Elter 1989. Hominoid dental variability and species number at the late Miocene site of Lufeng, China. *American Journal of Primatology* 18: 15-34.

Kelley, J. and D.R. Pilbeam 1986. The dryopithecines: taxonomy, anatomy and phylogeny of Miocene large hominoids. In *Comparative Primate Biology-Systematics, Evolution and Anatomy* (edited by D.R. Swindler and J. Erwin). New York: Alan R. Liss, Inc. 1: 361-411.

Kelley, J. and Q. Xu 1991. Extreme sexual dimorphism in a Miocene hominoid. *Nature* 352: 151-153.

Kennedy, K.A.R. 1984. Growth, nutrition, and pathology in changing palaeodemographic settings in South Asia. In *Palaeoanthropology at the Origins of Agriculture* (edited by M.N. Cohen and G.J. Armelagos). Orlando: Academic Press, Inc. pp. 169-192.

Kennedy, K.A.R. 1993. Recent discoveries and studies of the human skeletal remains of ancient Sri Lankans—a palaeoanthropological update. In *New Perspectives in Anthropology* (edited by P.K. Seth and S. Seth). New Delhi: MD Publications Pvt Ltd. pp. 299-341.

Kennedy, K.A.R. and S.U. Deraniyagala 1989. Fossil remains of 28,000 year old hominids in Sri Lanka. *Current Anthropology* 30: 394-399.

Kennedy, K.A.R. and Alison A. Elgart 1998. *Homind Remains: An Update—South Asia: India and Sri Lanka*. Belgium: Royal Belgium Institute of Natural Sciences.

Kennedy, K.A.R., A. Sonakia, J. Chiment and K.K. Verma 1991. Is the Narmada hominid an Indian *Homo erectus*? *American Journal of Physical Anthropology* 86:475-496.

Kerley, E.R. 1965. The microscopic determination of age in human bone. *American Journal of Physical Anthropology* 23: 149-164.

Kerley, E.R. and D.H. Ubelaker 1978. Revisions in the microscopic method of estimating age at death in human cortical bone. *American Journal of Physical Anthropology* 49: 545-546.

Khatri, A.P. 1961. Stone Age and Pleistocene chronology of the Narmada Valley, Central India. *Anthropos* 56: 519-530.

Khatri, A.P. 1975. The early fossil hominids and related apes of the Siwalik foothills of the Himalayas: recent discoveries and new interpretations. In *Palaeoanthropology-Morphology and Palaeoanthropology* (edited by R.H. Tuttle). The Hauge: Mouton Publishers. pp. 31-58.

Kimbel, W.H., D.C. Johanson and Y. Rak 1994. The first skull and other new discoveries of *Australopithecus afarensis* at Hadar, Ethiopia. *Nature* 368:449-451.

Kimbel, W.H. and L.B. Martin 1993. *Species, Species Concepts, and Primate Evolution.* New York: Plenum Press.

von Koenigswald, G.H.R. 1935. Eine fossile Saügetierfauna mit simia aus südchina. *Koniklijke Nederlands Akademie van Wetenschappen* 38: 872-879.

von Koenigswald, G.H.R. 1952. *Gigantopithecus blackii* von Koenigswald, a giant fossil hominoid from the Pleistocene of southern China. *Anthropological Papers of the American Museum of Natural History* 43: 292-325,

von Koenigswald, G.H.R. 1963. *The Evolution of Man.* Ann Arbor: The University of Michigan Press.

von Koenigswald, G.H.R. 1972. Ein Unterkiefer Eines fossilien Hominoiden aus dem Unterpliozän Griechenlands. *Proceedings of the Koninklijke Nederlandse Akademie von Wetenschappen,* Amsterdam series B, 75: 385-394.

von Koenigswald, G.H.R. 1976. *The Evolution of Man.* Ann Arbor: The University of Michigan Press.

Korn, N. 1973. *Human Evolution.* New York: Holt, Rinehart and Winston, Inc.

Krantz, G.S. 1968. Brain size and hunting ability in earliest man. *Current Anthropology* 9:450-451.

Krings, M., A. Stone, R.W. Schmitz, H. Kraintzki, M. Stoneking and S. Paabo 1997. *Neanderthal* DNA sequences and the origin of modern humans. *Cell* 90: 19-30.

Kroeber, A.L. 1939. *Cultural and natural areas of native North America.* Berkeley: University of California *Publications in American Archaeology and Ethnology* 38: 134.

Krynine, P.D. 1937. Petrography and genesis of the Siwalik series. *American Journal of Science* 34: 422-446.

Lallo, J.W., G.J. Armelagos and R.P. Mensforth 1977. The role of diet, disease and physiology in the origin of porotic hyperostosis. *Human Biology* 49:471-483.

Landau, M. 1984. Human evolution as narrative. *American Scientist* 72:262-268.

Larson, C.S. 1980a. Dental caries experimental and biocultural evidence. *Tennessee Anthropological Association Miscellaneous Papers* 5:75-80.

Larson, C.S. 1980b. Prehistoric human biological adaptation: case study from the Georgia Coast. Ph.D. Thesis (Unpublished). Ann Arbor: Department of Anthropology, University of Michigan.

Larson, C.S. 1981a. Functional implications of post-cranial size reduction on the prehistoric Georgia Coast, U.S.A. *Journal of Human Evolution* 10:489-502.

Larson, C.S. 1981b. Skeletal and dental adaptations to the shift to agriculture on the Georgia Coast. *Current Anthropology* 22:422-423.

Lartet, E. 1856. Note sur un grand Singe Fossile qui se rattache au groupe des Singes superieurs. *Comptes Rendus de l'Académie des Sciences* 42 (July 28).

Lartet, E. 1856. Note sur un grand singe fossile qui se rattache au groupe des singes supérieurs. *Comptes Rendus de l'Académie des Sciences* 43: 219.

Lartet, E. 1868. Une sépulture des troglodytes du Périgord. *Bull. de la Soc. d'Anthrop. de Paris,* III.

Laughlin, W.S. 1968. Hunting: an integrating biobehaviour system and its evolutionary importance. In *Man: The Hunter* (edited by R.B. Lee and I. de Vore). Chicago: Aldine. pp. 304-320.

Leakey, L.S.B. 1951. *Olduvai Gorge.* Cambridge: Cambridge University Press.

Leakey, L.S.B. 1959. A new fossil skull from Olduvai. *Nature* 184:491-493.

Leakey, L.S.B. 1961a. New finds at Olduvai Gorge. *Nature* 189:649-650.

Leakey, L.S.B. 1961b. The juvenile mandible from Olduvai. *Nature* 191:417-418.

Leakey, L.S.B. 1962. A new lower Pleistocene fossil primate from Kenya. *Ann. Mag. Nat. Hist. Ser.* 13(4):689-696.

Leakey, L.S.B. 1963. East African fossil hominoidea and the classification within this superfamily. In *Classification and Human Evolution* (edited by S.L. Washburn). Chicago: Aldine Publishing Company. pp. 32-49.

Leakey, L.S.B., J.F. Everden and G.H. Curtis 1961. Age of Bed I, Olduvai Gorge, Tanganyika. *Nature* 191:478-479.

Leakey, L.S.B., P.V. Tobias and J.R. Napier 1964. A new species of the genus *Homo* from Olduvai Gorge. *Nature* 202:7-9.

Leakey, R.E. 1981. *The Making of Mankind*. London: Michael Joseph Limited.

Leakey, R.E. and R. Lewin 1977. *Origins*. London: Macdonald and Jane's.

Leakey, R.E.F. 1972. Further evidence of Lower Pleistocene hominids from East Rudolf, North Kenya. *Nature* 237: 264-269.

Lengyel, I. 1968. Biochemical aspects of early skeletons. In *The Skeletal Biology of Earlier Populations* (edited by D.R. Brothwell). Oxford: Pergamon Press Ltd. pp. 271-288.

Lesthaeghe, R.I. and H.J. Page 1980. The post-partum non-susceptible period: development and application of model schedules. *Population Studies* 34:143-169.

Lévéque, F. and B. Vandermeersch 1980a. Les découvertes de restes humains dans un horizon castelperronien de Saint-Césaire (Charente-Maritime). *Comptes Rendus de l'Académie des Sciences* Paris Ser D 291:187-189.

Lévéque, F. and B. Vandermeersch 1980b. Les découvertes de restes humains dans un horizon castelperronien de Saint-Césaire (Charente-Maritime). *B. Bull. Soc. prehist. fr.* 77:35.

Lévéque, F. and B. Vandermeersch 1981. Le nêandertalien de Saint-Cêsaire. *La Recherche* 12:242-244.

Lewin, R. 1987a. *Bones of Contention—Controversies in the Search for Human Origins*. Penguin Books.

Lewin, R. 1987b. Africa: cradle of modern humans. *Science* 237:1292-1295.

Lewin, R. 1988a. Hip joints: clues to bipedalism. *Science* 241: 1433.

Lewin, R. 1988b. Conflict over DNA clock results. *Science* 241:1598-600; and 1756-1759.

Lewin, R. 1988c. Trees from genes and tongues. *Science* 242:514.

Lewin, R. 1988d. A new tool maker in the hominid record. *Science* 24:724-725.

Lewin, R. 1991. *Bones of Contention-Controversies in the Search for Human Origins.* London: Penguin Books.

Lewin, R. 1993a. *The Origin of Modern Humans.* New York: Scientific American Library.

Lewin, R. 1993b. *Human Evolution: An Illustrated Introduction*. 3rd ed. Boston: Blackwell Scientific Publications Inc.

Lewis, G.E. 1933. Preliminary notice of a new genus of Lemuroid from the Siwaliks. *American Journal of Science* 26: 134-138.

Lewis, G.E. 1936. A new species of *Sugrivapithecus*. American *Journal of Science* 31: 450-452.

Lewis, G.E. 1937. Taxonomic syllabus of Siwalik fossil anthropoids. *American Journal of Science* 34:137-147.

Lewis, G.E. 1934. Preliminary notice of new man-like apes from India: scientific research of the Yale India expedition. *American Journal of Science* 27:161-181.

Li, W.H. 1997. *Molecular Evolution*. Massachusetts: Sinnauer.

Livingstone, F.B. 1991 (1971). Phylogenies and the forces of evolution. *American Journal of Human Biology* 3: 83-89.

Long, J.C. 1993. Human molecular phylogenetics. *Annual Review of Anthropology* 22: 251-272.

Lovejoy, C.O. 1988. Evolution of human walking. *Scientific American* November 1988. pp. 82-89.

Lovejoy, O., R.S. Meindl, T.R. Pryzbeck, T.S. Barton, K.G. Heiple and D. Kotting 1977. Palaeodemography of the Libben site, Ottawa County, Ohio. *Science* 198: 291-293.

Lunt, D.A. 1969. An odontometric study of medieval Danes. *Acta Odontologica Scandinavica* 27, Suppl., 55: 1-73.

Lydekker, R. 1878. Notices of Siwalik mammals. Records of Geological *Survey of India,* Calcutta 11:64-104.

Lydekker, R. 1878. Notices of Siwalik mammals. Records of Geological *Survey of India,* 11: 64-85.

Lydekker, R. 1879. Notices of Siwalik mammals. Records of Geological *Survey of India,* 12: 33-52.

Lydekker, R. 1884. Rodents, ruminants and synopsis of Mammalia. *Memoirs of Geological Survey of India (Palaeontologia Indica 10:* 105), Series X 3: 105-134.

Lydekker, R. 1886. Siwalik Mammalia. Supplement one. *Memoirs of the Geological Survey of India* 10: 1-18.

MacInnes, D.G. 1943. Notes on the east African Miocene primates. *Journal of East Africa Uganda Natural History Society* 17:141-181.

Madden, C.T. and G.E. Lewis 1980. *Indopithecus giganteus* distinct from *Sivapithecus indicus. Primates* 21: 572-576.

Mai, L.L. 1983. A model of chromosome evolution and its bearing on cladogenesis in the Hominoidea. In *New Interpretations of Ape and Human Ancestory* (edited by R.L. Ciochon and R.S. Corruccini). New York: Plenum Publishing Corporation. pp. 87-114.

Martin, D.L., G.J. Armelagos, A.H. Goodman and D.P. van Gerven 1984. The effects of socioeconomic change in prehistoric Africa: Sudanese Nubia as a case study. In *Palaeoanthropology at the Origins of Agriculture* (edited by M.N. Cohen and G.J. Armelagos). Orlando: Academic Press Inc. pp. 193-214.

Martin, L.B. and P. Andrews 1993. Species recognition in Middle Miocene hominoids. In *Species, Species Concepts, and Primate Evolutions* (edited by W.H. Kimbel and L.B. Martin). New York: Plenum Press. pp. 393-427.

Martin R.D. 1990. *Primate Origins and Evolution: A Phylogenetic Reconstruction*. London: Chapman and Hall.

Masset, C. 1971. *Erreurs system antiques dans la determination de l'age par les sutures craniennes. Bulletin et Memoires de la Societe d'Anthropologie de Paris,* Ser. XII.

Masters, P.M. and M.R. Zimmerman 1978. Age determination of an Alaskan mummy: morphological and biochemical correlation. *Science* 201: 811-812.

Mayr, E. 1974. Cladistic analysis or cladistic classification? *Zool. Syst. Evol. -Forsch.* 12:94-128.

McArthur, A.J. and J.A. Clark 1987. Body temperature and heat and water balance. *Nature* 326: 647-648.

McDermott, F., R. Grün, C.B. Stringer and C.J. Hawkesworth 1993. Mass spectrophotometric U-series dates for Israeli *Neanderthal/* early modern hominid sites. *Nature* 363:252-255.

McHenry, H. 1985. Implications of post-canine megadontia for the origin of *Homo*. In *Ancestors: The Hard Evidence* (edited by E. Delson). New York: Alan R. Liss Inc. pp. 178-183.

McHenry, H.M. and L.R. Berger 1998. Body proportions of *Australopithecus afarensis* and *A. africanus* and the origin of the genus *Homo*. *Journal of Human Evolution* 35: 1-22.

McKie, R. 1992. Genes rock the fossil record. *Geographical Magazine* November 1992. pp. 20-25.

Meindl, R.S. and C.O. Lovejoy 1983. Estimation of adult age at death based on the lateral-anterior sutures: methods and limits of inference in palaeodemography. *American Journal of Physical Anthropology* 60: 225.

Meindl, R.S., C.O. Lovejoy and R.P. Mensforth 1980. Multifactorial determination of skeletal populations of known age. *American Journal of Physical Anthropology* 52: 255

Mellars, P. 1989. Major issues in the emergence of modern humans. *Current Anthropology* 30:349-385.

Menaker, L. 1980. *Biologic Basis of Dental Caries: An Oral Biology Text Book.* New York: Harper and Row.

von Meyer, H. 1848. *Index Palaeontologiscus.* (quoted from Colbert, 1935).

Miller, D.A. 1977. Evolution of primate chromosomes. *Science* 198: 1116-1124.

Miller, D.A., I.L. Firschein, V.G. Dev, R. Tantravahi and O.J. Miller 1974. The gorilla karyotype: chromosome lengths and polymorphisms. *Cytogenetics and Cell Genetics* 13: 536-550.

Miller, J.A. 1991. Does brain size variability provide evidence of multiple species in *Homo habilis*? *American Journal of Physical Anthropology* 84:385-398.

Mittre, V. 1979. Environment of early man in North-West India: palaeobotanical evidence. In *Early Man in North West India* (edited by S.R.K. Chopra). New Delhi: Allied Publishers Private Limited. pp. 20-65.

Molnar, S. 1972. Tooth wear and culture: a survey of tooth functions among some prehistoric populations. *Current Anthropology* 13:511-526.

Molnar, S. 1983. *Human Variation*. Englewood Cliffs, New Jersey: Prentice-Hall Inc.

Montagu, M.F.A. 1957. The chin of the Kanam mandible. *American Anthropologist* 59: 335-338.

Montagu, M.F.A. 1945. *An Introduction to Physical Anthropology*. Springfield, Illinois: C.C. Thomas.

Moodie, R.L. 1923. *Palaeoanthropology*. An introduction to the study of ancient evidences of disease. Urbana: University of Illinois Press.

Mooney, J. 1928. *The Aboriginal Population of America North of Mexico*, edited by J.R. Swanton. *Smithsonian Miscellaneous Collections* 80:1-40.

Morrison, N.A., J.C. qi, A. Tokita, P.J. Kelly, L. Crofts, T.V. Nguyen, P.N. Sambrook and J.A. Eisman 1994. Prediction of bone density from vitamin D receptor alleles. *Nature* 367:284-287.

Moss, M.L. and L. Moss-Salentijn 1977. Analysis of developmental processes possibly related to human dental sexual dimorphism in permanent and deciduous canines. *American Journal of Physical Anthropology* 46: 407-414.

Mukerji, A.B. 1979. The early man's habitat in the Sivalik Hills: some statements on the geomorphological setting. In *Early Man in North West India* (edited by S.R.K. Chopra). New Delhi: Allied Publishers Private Limited. pp. 98-106.

Mundy, G.R. 1994. Boning up on genes. *Nature* 367:216-217.

Munthe, J., B. Dongol, J.H. Hutchison, W.F. Kean, K. Munthe and R.M. West 1983. New fossil discoveries from the Miocene of Nepal include a hominoid. *Nature* 303 (No. 5915): 331-333.

Murril, R. 1983. On the dating of the fossil hominid Petralona skull. *Anthropos* 10:12-15.

Napier, J. 1962a. The evolution of the human hand. *Scientific American* 207:56-62.

Napier, J. 1962b. Fossil hand bones from Olduvai Gorge. *Nature* 196:409-411.

Nei, M. 1973. Analysis of gene diversity in subdivided populations. *Proceedings of the National Academy of Sciences,* U.S.A. 70:3321-3323.

Nei, M. 1987. *Molecular Evolutionary Genetics.* New York: Columbia University Press.

Nei, M. 1996. Phylogenetic analysis in molecular evolutionary genetics. *Annual Review of Genetics* 30:371-403.

Nei, M. and G. Livshits 1989. Genetic relationships of Europeans, Asians and Africans and the origin of *Homo sapiens. Human Heredity* 39:276-281.

Nelson, H. and R. Jurmain 1988. *Introduction to Physical Anthropology.* New York: West Publishing Company.

Oldham, T. 1868. On the agate-flake found by Mr. Wynne in the Pleistocene (?) deposits of the Upper Godavery. *Records of the Geological Survey of India* 1: 65-69.

Ortner, D.J. 1992. Skeletal paleopathology: probabilities, possibilities and impossibilities. In *Disease and Demography in the Americas* (edited by J.W. Verano and D.H. Ubelaker). Washington and London: Smithsonian Institution Press. pp. 5-13.

Ortner, D.J. and A.C. Aufderheide 1991. *Human Palaeopathology: Current Synthesis and Future Options.* Washington and London: Smithsonian Institution Press.

Owsley, D.W. 1981. Dental discriminant sexing of Arikara skeletons.

Plains Anthropology 26: 165-169.

Owsley, D.W. 1983. Human bones from archaeological context: an important source of information. *Tennessee Anthropologist* 8: 20-27.

Oxnard, C.E. 1969. Evolution of the human shoulder: some possible pathways. *American Journal of Physical Anthropology* 30: 319.

Oxnard, C.E. 1975. The place of the australopithecines in human evolution: grounds for doubt? *Nature* 258:389-395.

Oxnard, C.E. 1984. *The Order of Man: A Biomathematical Anatomy of the Primates*. New Haven, CT: Yale University Press.

Oxnard, C.E. 1987. *Fossils, Teeth and Sex: New Perspectives on Human Evolution*. Seattle and London: University of Washington Press.

Pandey, J. and V.V. Sastry 1968. On a new species of *Sivapithecus* from the Siwalik rocks of India. *Journal of Geological Society of India* 9: 206-211.

Peebles, C.S. 1977. Biocultural adaptation in prehistoric America: an archaeologist's perspective. In *Biocultural Adaptation in Prehistoric America* (edited by R.L. Blakely). *Southern Anthropological Society Procs.* No. 11:115-130. Athens: The University of Georgia Press.

Perrin-Pecontal, P., M. Gouy, V.-M. Nignon and G. Trabuchet 1992. Evolution of the primate β-globin region: Nucleotide sequence of the β-globin intergenic region of Gorilla and phylogenetic relationships between African apes and man. *Journal of Molecular Evolution* 34: 17-30.

Perzigian, A.J., P.A. Tench and D.J. Braun 1984. Prehistoric health in the Ohio River Valley. In *Palaeopathology at the Origins of Agriculture* (edited by M.N. Cohen and G.J. Armelagos). Orlando: Academic Press, Inc. pp. 347-366.

Pfeiffer, J.E. 1969. *The Emergence of Man*. New York/London: Harper & Row Publishers.

Phenice, T.W. 1969. A newly developed visual method of sexing of pubis. *American Journal of Physical Anthropology* 30:297-302.

Pilbeam, D. 1970. *The Evolution of Man*. London: Thames and Hudson.

Pilbeam, D. 1985. Patterns of hominoid evolution. In *Ancestors: The Hard Evidence* (edited by E. Delson). New York: Alan R. Liss. pp. 51-59.

Pilbeam, D. 1996. Personal Communication.

Pilbeam, D., G.E. Meyer, C. Badgley, M.D. Rose, M.H.L. Pickford, A.K.

Behrensmeyer and S.M.I. Shah 1977 New hominoid primates from the Siwaliks of Pakistan and their bearing on hominoid evolution. *Nature* 270: 689-695.

Pilbeam, D.R. 1969. Newly recognised mandible of *Ramapithecus. Nature* (Lond.) 222: 1093-1094.

Pilbeam, D.R. 1970. *Gigantopithecus* and the origins of the Hominidae. *Nature* (Lond.) 225: 516-519.

Pilbeam, D.R. 1982. New hominoid skull material from the Miocene of Pakistan. *Nature* (Lond.) 295: 232-234.

Pilbeam, D.R., M. Rose, C. Badgley and B. Lipschutz 1980. Miocene hominoids from Pakistan. *Postilla* 181: 1-94.

Pilbeam, D.R. and E.L. Simons 1971 Humerus of *Dryopithecus* from Saint Gaudens, France. *Nature* 229:406-407.

Pilgrim, G.E. 1910. Notices of a new mammalian genera and species from the Tertiaries of India. *Records of Geological Survey of India* 40: 63-71.

Pilgrim, G.E. 1915. New Siwalik primates and their bearing on the question of the evolution of man and the Anthropoidea. *Records of Geological Survey of India* 45:1-74.

Pilgrim, G.E. 1927. A *Sivapithecus* palate and other primate fossils from India. *Memoirs of Geological Survey of India (Palaeontologia Indica),* n.s. 14:1-26.

Pilgrim, G.E. 1932. The fossil carnivora of India. *Palaeontologia Indica,* n.s., 18: 1-232.

Poirier, F.E. 1987. *Understanding Human Evolution.* New Jersey: Prentice Hall Inc.

Poirier, F.E., W.A. Stini and K.B. Wreden 1990. *In Search of Ourselves—An Introduction to Physical Anthropology.* New Jersey: Prentice Hall.

Potts, R. 1984. Home base and early hominids. *American Scientist* 72:338-347.

Potts, R. 1986. Temporal span of bone accumulations at Olduvai Gorge and implications for early hominid foraging behaviour. *Palaeobiology* 12:25-31.

Powell, M.L. 1985. The analysis of dental wear and caries for dietary reconstruction. In *The Analysis of Prehistoric Diets* (edited by R.I. Gilbert, Jr. and J.H. Mielke). Orlando: Academic Press Inc. pp 307-338.

Prakash, P.V. 1990. Prehistoric/palaeoanthropological research in contemporary India. Paper presented at the Special Symposium on *Anthropology in Contemporary India* Patiala January 8-9, 1990 (Jointly organised by the Department of Human Biology, Punjabi University, Patiala and the Anthropological Survey of India on the occasion of its 150th year of ASI, Calcutta).

Prasad, K.N. 1962. Fossil primates from the Siwalik beds near Haritalyangar, Himachal Pradesh, India. *Journal of the Geological Society of India* 3: 86-96.

Prasad, K.N. 1964. Upper Miocene anthropoids from the Siwalik beds of Haritalyangar, Himachal Pradesh, India. *Paleontology* 7: 124.

Prasad, K.N. 1968. The vertebrate fauna from the Siwalik beds of Haritalyangar, Himachal Pradesh, India. *Palaeontologia Indica* 39: 1-55.

Prasad, K.N. 1969a. Fossil anthropoids from the Siwalik system of India. In *Recent Advances in Primatology* (edited by H.O. Hofer). Basel and New York: S. Karger. 2: 131-134.

Prasad, K.N. 1969b. Critical observations on the fossil anthropoids from the Siwalik system of India. *Folia Primatologica* 10: 228-317.

Prasad, K.N. 1969c. Observations of mid-Tertiary hominids *Sivapithecus* and *Ramapithecus*. *American Journal of Physical Anthropology* 31:11.

Prasad, K.N. 1970. The vertebrate fauna from the Siwalik beds of Haritalyangar, Himachal Pradesh, India. *Palaeontologia Indica,* n.s. 39: 1-79.

Prasad, K.N. 1971. Ecology of the fossil *Hominoidea* from the Siwaliks of India. *Nature* (Lond.) 232: 413-414.

Putschar, W. 1931. *Entwicklung, Wachstum Und Pathologie Des Beckenverbindungen Des Menschen.* Jens: Gustav Fischer Verlag.

Rak, Y. 1985. Australopithecinae taxonomy and phylogeny in light of facial morphology. *American Journal of Physical Anthropology* 66:281-288.

Rathbun, T.A. 1984. Skeletal pathology from the palaeolithic through the metal ages in Iran and Iraq. In *Palaeopathology at the Origins of Agriculture* (edited by M.N. Cohen and G.J. Armelagos). Orlando: Academic Press Inc. pp. 137-167.

Rathbun, T.A., J. Sexton and J. Michie 1980. Disease patterns in a formative period, South Carolina Coastal Population. *Tennessee Anthropological Association Miscellaneous Papers* 5:52-74.

Raza, S.M., J.C. Barry, D. Pilbeam, M.D. Rose, S.M. Ibrahim Shah and S. Ward 1983. New hominoid primates from the middle Miocene Chinji formation, Potwar Plateau, Pakistan. *Nature* 306: 52-54.

Remane, A. 1956. Paläontologie und Evolution der Primaten.In

Primatologia (edited by H. Hofer, A.H. Schultz and D. Starck). Basel: S. Karger. Vol. 1, pp. 267.

Rightmire, G.P. 1976. Relationships of middle and upper Pleistocene hominids from Sub-Saharan Africa. *Nature* 269:238-240.

Rightmire, G.P. 1979. Cranial remains of *Homo erectus* from Beds II and IV, Olduvai Gorge, Tanzania. *American Journal of Physical Anthropology* 51:99-116.

Rightmire, P. 1984 (1883). Comparison of *Homo erectus* from Africa and South-East Asia. Cours. *Forschungs. Senckenburg* 69: 83-98.

Robinson, J. 1963a. Adaptive radiation in the australopithecines and the origin of man. In *African Ecology and Human Evolution* (edited by F. Howell and F. Bourliere). Chicago: Aldine Publishing Company. pp. 385-416.

Robinson, J. 1963b. Australopithecines, culture and phylogeny. *American Journal of Physical Anthropology* 21:595.

Robinson, J.T. 1952. Some hominid features of the ape-man dentition. *Journal of Dental Association of South Africa* 7: 102.

Robinson, J.T. 1956. The dentition of the australopithecinae. *Mem. Transvaal Mus.* 9:1-179.

Rose, J.C., K.W. Condon and A.H. Goodman 1985. Diet and dentition: developmental disturbances. In *The Analysis of Prehistoric Diets* (edited by R.I. Gilbert, Jr. and J.H. Mielke). Orlando: Academic Press Inc. pp. 307-338.

Ruvolo, M. and D.R. Pilbeam 1986. Hominoid evolution: Molecular and palaeontological pattern. In *Major Topics in Primate and Human Evolution* (edited by B.A. Wood, L. Martin and P. Andrews). Cambridge: Cambridge University Press. pp. 157-160.

Sahni, A., V. Kumar and V.C. Srivastava 1974. *Dryopithecus* (subgenus: *Sivapithecus*) and associated vertebrates from the lower Siwaliks of Uttar Pradesh. *Bull. Indian Geological Association* 7: 54.

Sahni, M.R. 1954. A century of palaeontology, palaeobotany and prehistory in India and adjacent countries. *Journal of the Palaeontological Society of India* 1: 7-51.

Sandison, A.T. 1968. Pathological changes in the skeletons of earlier populations due to acquired disease and difficulties in their interpretation. In *The Skeletal Biology of Earlier Human Populations* (edited by D.R. Brothwell). Oxford: Pergamon Press Ltd. pp. 205-243.

Sankhyan, A.R. 1986. A fresh look on the fossil hominoidea and the problem of last common ancestry. *Procs. Indian Science Congress*

Association.

Sarich, V. 1971. A molecular approach to the question of human origins. In *Background for Man: Readings in Physical Anthropology* (edited by P. Dolhinow and V.M. Sarich). Boston: Little Brown. pp. 60-81.

Sarich, V. and A. Wilson 1967a. Immunological time scale for hominid evolution. *Science* 158:1200-1203.

Sarich, V. and A. Wilson 1967b. Rates of albumin evolution in primates. *Procs. National Academy of Sciences,* U.S.A. 58:142-148.

Saxena, M.N. 1976. The Himalayan uplift, antecedent drainage and the elusive fossil-man in north-western India. *Recent Researches in Geology* 8: 139-152.

Schaller, G.B. and G.R. Lowther 1969. The relevance of carnivore behaviour to the study of early hominids. *South-western Journal of Anthropology* 25: 325-336.

Schlosser, M. 1887. Die Affen, Lemuren, Chiropteren etc des europäischen Tertiärs. *Beitr. Paläont. Öst-Ungarns 6* (cited from Hürzeler, 1958).

Schlosser, M. 1903. Die fossielen säugetiere Chinas. *Abhandlungen der Bayerische Akademie der Wissenschaften* 22:1-221.

Schlosser, M. 1910. Uber einige fossile Säugetiere aus dem Oligocän von Aegypten. *Zoologischer Anzeiger* 35:500.

Schlosser, M. 1911a. Beitrage zur Kenntnis der oligozänen Landsäugetiere aus dem Fayum. *L'Anthropologie* 23:417.

Schlosser, M. 1911b. Ägypten. *Beitrage Palaeont. Geol. Österreich-Ungarns u.d. Orients* 24(6):1-227.

Schrenk, F., T.G. Bromage, C.G. Betzler, U. Ring and Y.M. Juwayeyi 1993. Oldest *Homo* and Pliocene biogeography of the Malawi Rift. *Nature* 365:833-836.

Schultz, A.H. 1967. Notes on diseases and healed fractures of wild apes. In *Diseases in Antiquity* (edited by D.R. Brothwell and A.T. Sandison). Springfield, Illinois: C.C. Thomas. pp. 47-55.

Schultz, M. 1981. Asceptic bones necrosis found in the skeletal material from Merovingian cemetery of Kleinjangheim, Southern Germany. *Palaeopathology Newsletter* 34:7-9.

Schwalbe, G. 1915. Über den fossilen Affen *Oreopithecus bambolii*: Zugleich ein Beitrag zur Morphologie der Zähne der Primaten. *Zeitschrift fur Morphologie und Anthropologie* 19:149-254.

Schwartz, A.H. 1990. *Lufengpithecus* and its potential relationship to an orangutan clade. *Journal of Human Evolution* 19: 591-605.

Schwartz, J.H. and I. Tattersall 1996. Whose teeth? *Nature* 381: 201-202.

Seth, P.K. and S. Seth 1986. A review of evolutionary and genetic differentiation in primates. In *Primate Evolution* (edited by Else and Lee). Cambridge: Cambridge University Press. *Procs. 10th Congress of the International Primatological Society*. Vol. 1, pp. 291-306.

Seth, P.K. 1975. Evolutionary dynamics of chromosomal genetic material. *Journal of Cytology and Genetics Congr. Suppl.* 165-172.

Seth, P.K. and M.B. Saxena 1977. Cytological mechanisms in the evolution of mammalian chromosomes. *The Nucleus* 20:42-52.

Seth, P.K., L.E.M. de Boer, M.B. Saxena and S. Seth 1976. Comparison of human and non-human primate chromosomes using the fluorescent benzimidazol and other banding techniques. *Chromosomes Today* 5: 315-322.

Shea, B.T., S.R. Leogh and C.P. Groves 1993. Multivariate craniometric variation in chimpanzees—implications for species identification in palaeoanthropology. In *Species, Species Concepts, and Primate Evolution* (edited by W.H. Kimbel and L.B. Martin). New York: Plenum Press. pp. 265-296.

Shipman, P. 1986. Baffling limb on the family tree. *Discover* September issue pp. 87-93.

Shipman, P. 1993. On the origin of races. *New Scientist* 16th January, pp. 34-37.

Sibley, C.G. and J.E. Ahlquist 1984. The phylogeny of the hominoid primates as indicated by DNA-DNA hybridization. *Journal of Molecular Evolution* 20:2-15.

Sillen, A. and P. Smith 1983. Strontium/calcium ratios reveal weaning-age in a skeletal population. *American Journal of Physical Anthropology* 60:253.

Simons, E.L. 1960a. New fossil primates: a review of the past decade. *American Scientist* 48:179-192.

Simons, E.L. 1960b. *Apidium* and *Oreopithecus*. *Nature* 186:824-826.

Simons, E.L. 1961. The phyletic position of *Ramapithecus*. *Postilla* 57:1-9.

Simons, E.L. 1964. The early relatives of man. *Scientific American*.

Simons, E.L. 1969. The origin and radiation of the primates. *Annals of New York Academy of Sciences* 167:319-331.

Simons, E.L. 1972. *Primate Evolution—an introduction to man's place in nature*. New York: Macmillan.

Simons, E.L. 1974. On the discovery of *Gigantopithecus* in north India. In *Perspectives in Palaeoanthropology* (edited by A.K. Ghosh). Calcutta: Firma K.L. Mukhopadhyay. pp. 1-7.

Simons, E.L. 1976a. The nature of the transition in the dental mechanism from pongids to hominids. *Journal of Human Evolution* 5:511-528.

Simons, E.L. 1976b. The fossil record of primate phylogeny. In *Molecular Anthropology* (edited by M. Goodman, R.E. Tashian and J.H. Tashian). New York: Plenum Publishing Corporation. pp. 35-62.

Simons, E.L. 1977. *Ramapithecus. Scientific American* 236(5):28-35.

Simons E.L. and S.R.K. Chopra 1969a. A preliminary announcement of a new *Gigantopithecus* species from India. In *Recent Advances in Primatology* (edited by H.O. Hofer). Basel and New York: S. Karger. 2: 135-142.

Simons, E.L. and S.R.K. Chopra 1969b. *Gigantopithecus (Pongidae, Hominoidea)*: a new species from North India. Postilla (Yale University Peabody Museum of Natural History) 138:1-18.

Simons, E.L. and D. Pilbeam 1965. Preliminary revision of the dryopithecinae (Pongidae, Anthropoidea). *Folia Primatologia* 3:81-152.

Simons, E.L. and D.R. Pilbeam 1972. Hominid palaeoprimatology. In *The Functional and Evolutionary Biology of Primates* (edited by R.H. Tuttle). Chicago: Aldine. pp. 36-62.

Simons, E.L. and D.R. Pilbeam 1978. *Ramapithecus*. In *Evolution of African Mammals* (Edited by V.J. Maglio and H.B.S. Cooke). Massachusetts: Harvard University Press. pp. 147-153.

Simpson, G.G. 1935. The first mammals. *Quarterly Review of Biology* 10:154-180.

Simpson, G.G. 1945. The principles of classification and a classification of mammals. *Bulletin of American Museum of Natural History* 85:1-350.

Simpson, G.G. 1949a. Rates of evolution in animals. In *Genetics, Palaeoantology and Evolution* (edited by G.L. Jepsen et al.). Princeton: University Press. pp. 205-228.

Simpson, G.G. 1949b. *The Meaning of Evolution*. New Haven: Yale University Press.

Sjovold, T. 1975. Tables of the combined method for determination of age at death given by Nemeskeri, Harsanyi and Acsadi. *Anthropologiai Kozlemenyek* 19:9-22.

Smith, F.H. 1984. Fossil hominids from the upper Pleistocene of central Europe and the origin of modern Europeans. In *The Origins of*

Modern Humans (edited by F.H. Smith and F. Spencer). New York: Alan R. Liss. pp. 137-209.

Smith, P., O. Bar-Yosef and Sillen 1984. Archaeological and skeletal evidence for dietary change during the late Pleistocene/early Holocene in Levant. In *Palaeopathology at the Origins of Agriculture* (edited by M.N. Cohen and G.J. Armelagos). Orlando: Academic Press, Inc. pp. 101-136.

Sola, S.M. and M. Köhler 1993. Recent discoveries of *Dryopithecus* shed new light on evolution of great apes. *Nature* 365:543-545.

Sonakia, A. 1984. The skull cap of early man and associated mammalian fauna from Narmada Valley alluvium, Hoshangabad area, Madhya Pradesh (India). *Records of the Geological Survey of India* 113(6):159-172.

Sonakia, A. 1984. The skull cap of early Man and associated mammalian fauna from the Narmada Valley Alluvium, Hoshangabad Area, Madhya Pradesh, India. *Records of the Geological Survey of India* 113: 159-172.

Sonakia, A. 1985a. Early *Homo* from Narmada Valley, India. In *Ancestors: The Hard Evidence* (edited by E. Delson). New York: Alan R. Liss, Inc. pp. 334-339.

Sonakia, A. 1985b. Skull cap of an early Man from the Narmada Valley Alluvium (Pleistocene) of Central India. *American Anthropologist* 87: 612-616.

Stones, H.H. 1954. *Oral and Dental Diseases*. Edinburgh: Livingstone.

Stout, S.D. and S.J. Gehlert 1980. The relative accuracy and reliability of histological aging methods. *Forensic Science International* 15: 181-190.

Stout, S.D. 1988. The use of histomorphology to estimate age. *Journal of Forensic Sciences* 33: 121-125.

Straus, W.L. Jr. 1949. The riddle of man's ancestry. *Quarterly Review of Biology* 24:200-223.

Straus, W.L., Jr. 1963. The classification of *Oreopithecus.* In *Classification and Human Evolution* (edited by S.L. Washburn). Chicago: Aldine. pp. 146-177.

Straus, W.L., Jr. and A.J.E. Cave 1957. Pathology and the posture of *Neanderthal* man. *Quarterly Review of Biology* 32: 348-363.

Stringer, C.B. 1974. Population relationships of later Pleistocene hominids: a multivariate study of available crania. *Journal of Archaeological Science* 1:317-342.

Stringer, C. 1984a. Human evolution and biological adaptation in the

Pleistocene. In *Hominid Evolution and Community Ecology* (edited by R. Foley). London: Academic Press, Inc. pp 55-83.

Stringer, C. 1984b. Fate of the *Neanderthal*. *Natural History* December: 6-12.

Stringer, C. 1990. The Asian connection. *New Scientist* 17th November. pp. 33-37.

Stringer, C. 1993. Secrets of the pit of the bones. *Nature* 362:501-502.

Stringer, C.B., J.J. Hublin and B. Vandermeersch 1984. The origin of anatomically modern humans in Western Europe. In *The Origins of Modern Humans* (edited by F.H. Smith and F. Spencer). New York: Alan R. Liss. pp. 51-135.

Susman, R.L. 1994. Fossil evidence for early hominid tool use. *Science* 265: 1570-1573.

Suwa, G., B. Asfaw, Y. Beyene, T.D. White, S. Katoh, S. Nagaoka et al. 1997. The first skull of *Australopithecus boisei*. *Nature* 389: 489-492.

Suzuki, H. and F. Takai 1970. *The Amud Man and His Cave Site*. Tokyo: Keigaku Publishing Co.

Swisher III, C.C., W.J. Rink, S.C. Anton, H.P. Schwartz, G. Curtis, A. Supryo and Widiasmoro 1996. Latest *Homo erectus* of Java: potential contemporaneity with *Homo sapiens* in southeast Asia. *Science* 274: 1870-1874.

Szalay, F.S. and E. Delson 1979. *Evolutionary History of the Primates.* London and New York: Academic Press Inc.

Tanner, J.M. 1962. *Growth at adolescence.* 2nd edition. Oxford: Blackwell Scientific Publications.

Tardieu, C. 1981. Morpho-functional analysis of the articular surfaces of the knee-joint in primates. In *Primate Evolutionary Biology* (edited by B. Chiarelli and R. Corruccini). Berlin: Springer-Verlag. pp. 68-80.

Tardieu, C. 1983. *L'Articulation du Genou. Analyse morphofunctionelle chez les Primates et les Hommes fossiles.* Paris: Editions CNRS.

Tattersall, I. 1968. A mandible of *Indraloris* (Primates, *Lorisidae*) from Miocene of India. *Postilla* 123: 1-10.

Tattersall, I. 1969. More on the ecology of *Ramapithecus*. *Nature* (Lond.) 224: 821-822.

Tattersall, I. 1993. *The Human Odyssey: Four Million Years of Human Evolution.* New York: Prentice Hall.

Tattersall, I. 1995. *The Fossial Trail.* New York: Oxford University

Press.

Tattersall, I. *et al.* 1988. *Encyclopaedia of Evolution and Prehistory.* New York and London: Garland.

Tattersall, I. and J.H. Schwartz 1996. Significance of some previously unrecognized apomorphies in the nasal region of *Homo neanderthalensis. Procs. National Academy of Sciences,* U.S.A. 93: 10852-10854.

Templeton, A.R. 1992. Human origins and analysis of mitochondrial DNA sequences. *Science* 235:737.

Theobald, W. 1880. On the Tertiary and alluvial deposits of central portion of the Narmada valley. *Memoirs of the Geological Survey of India* 2: 279-298.

Theobald, W. 1881. The Siwalik group of the sub-Himalayan region. *Records of the Geological Survey of India* 14: 66-125.

Thomas, H. 1985. The early and middle Miocene land connection of the Afro-Arabian Plate and Asia: a major event for hominoid dispersal? In *Ancestors: The Hard Evidence* (edited by E. Delson). New York: Alan R. Liss, Inc. pp. 42-50.

Thorne, A.G. and M.H. Wolpoff 1992. The multiregional evolution of humans. *Scientific American* April, pp. 76-83.

Tianyuan, L. and D.A. Etler 1992. New middle Pleistocene hominid crania from Yunxian in China. *Nature* 357:404-407.

Tobias, P.V. 1964. The Olduvai Bed I hominine with special references to its cranial capacity. *Nature* 203:3.

Tobias, P.V. 1965. Early man in east Africa. *Science* 149:22-33.

Tobias, P.V. 1969. Commentary on new discoveries and interpretations of early African fossil hominids. *Yearbook of Physical Anthropology,* pp. 24-30.

Tobias, P.V. 1971. *The Brain in Hominid Evolution.* New York: Columbia University Press.

Tobias, P.V. 1993. Earliest *Homo* not proven. *Nature* 361:307.

Tobias, P.V. 1994. The evolution of early hominids. In *Companion Encyclopaedia of Anthropology* (edited by T. Ingold). London and New York: Routledge. pp. 33-78.

Todd, N.B. and B. Blumenberg 1974. On the adaptive radiation of hominids. *Current Anthropology* 15: 383-385.

Trinkaus, E. 1980. Sexual differences in *Neanderthal* limb bones. *Journal of Human Evolution* 9:377-397.

Trinkaus, E. 1981. Neandertal limb proportions and cold adaptation.

In *Aspects of Evolution* (edited by C.B. Stringer). London: Taylor and Francis.

Trinkaus, E. 1983. *The Shanidar Neanderthals.* New York: Academic Press.

Trinkaus, E. 1984. Western Asia. In *The Origins of Modern Humans* (edited by F.H. Smith and F. Spencer). New York: Alan R. Liss.

Trinkaus, E. 1985. Pathology and posture of the *La Chapelle aux Saints Neandertal. American Journal of Physical Anthropology* 67:19-41.

Trinkaus, E. 1992. Evolutionary continuity among archaic *Homo sapiens*. In *Transition from Lower to Middle Palaeolithic and the Origin of Modern Man* (edited by A. Ronen). Oxford: BAR International Series 151.

Trinkaus, E. 1991. The evolution and dispersal of modern humans in Asia. *Current Anthropology* 32:353-355.

Turleau, C., N. Creau-Goldberg, C. Cochet and J. de Grouchy 1983. Gene mapping of the gibbon. Its position in primate evolution. *Human Genetics* 64: 65-72.

Tuttle, R.H. 1967. Knuckle-walking and the evolution of hominoid hands. *American Journal of Physical Anthropology* 26:171-206.

Tuttle, R.H. 1969. Knuckle walking and the problem of human origins. *Science* 166: 953.

Ubelaker, D.H. 1974. *Reconstruction of demographic profiles from ossuary skeletal samples. A case study from the tidewater Potomac.* Smithsonian Contributions to Anthropology, No. 18. Washington, D.C.: Smithsonian Institution Press.

Ubelaker, D.H. 1978. *Human Skeletal Remains.* Chicago: Aldine Publications.

Ubelaker, D.H. 1988. Human remains from OGSE – 46, La Libertad, Guayas Province, Ecuador. *Journal of Washington Academy of Sciences* 78(1):1-37.

Ubelaker, D.H. Prehistoric new world population size: historical review and current appraisal of North American estimates. *American Journal of Physical Anthropology* : 661-665.

Vallois, H.V. 1946. *Les Hommes Fossiles*. Paris: Masson et C[ie].

Vallois, H.V. 1954. *Neanderthals* and presapiens. *Journal of the Royal Anthropological Institute* 84:111-130.

van Valen, L. and R.E. Sloan 1965. The earliest primates. *Science* 150:743-745.

Vasishat, R.N. 1985. *Antecedents of Early Man in North-Western India*

Palaeontological and Palaeoecological Evidences. New Delhi: Inter-India Publications.

Vasishat, R.N., R. Gaur and S.R.K. Chopra 1978a. Community structure of middle Sivalik vertebrates from Haritalyangar (H.P.), India. *Palaeogeography, Palaeoclimatology, Palaeoecology* 23: 131-140.

Vasishat, R.N., R. Gaur and S.R.K. Chopra 1978b. Geology, fauna and palaeoenvironments of lower Sivalik deposits around Ramnagar, India. *Nature* 275: 736-737.

Verma, B.C. 1968. *Procynocephalus pinjorii* sp. Nov. A new fossil primate from Pinjore beds (Lower Pleistocene) east of Chandigarh. *Journal of Palaeontological Society of India* 13: 53-57.

Viegas-Pequignot, E., J. Couturier and B. Dutrillaux 1978. Comparison of DNA replication chronology in chromosomes of chimpanzee and man. *Primates* 19: 209-213.

Vigilant, L.R., R. Pennington, H. Harpending, T.D. Kocher and A.C. Wilson 1989. Mitochondrial DNA sequences in single hairs from southern African populations. *Proc. National Academy of Sciences,* USA 86:9350-9354.

Vigilant, L.R., M. Stoneking, H. Harpending, K. Hawkes and A.C. Wilson 1991. African populations and the evolution of mitochondrial DNA. *Science* 253:1503-1507.

Vogel, C. 1975. Remarks on the reconstruction of the dental arcade of *Ramapithecus*. In *Palaeoanthropology-Morphology and Palaeoecology* (edited by R.H. Tuttle). The Hauge: Mouton Publishers. pp. 87-98.

Vosberg, H.-P. 1989. The polymerase chain reaction: an improved method for the analysis of nucleic acids. *Human Genetics* 83:1-15.

W.E. Le Gros 1955. *The Fossil Evidence for Human Evolution*. Chicago: University Press.

Waddle, D.M. 1994. Matrix correlation tests support a single origin for modern humans. *Nature* 368:452-454.

Wadia, D.A. and N.K.N. Aiyengar 1938. Fossil anthropoids from India: a list of the Tertiary deposits of India. *Records of Geological Survey of India* 72: 467-494.

Wagner, W.H. 1969. The construction of a classification. In *Systematic Biology*. Washington, D.C.: National Academy of Sciences Publication No. 1692, pp. 67-99.

Walker, A. and R.E.F. Leakey 1979. The hominids of east Turkana. In *Human Ancestors* (edited by G. Isaac and R.E.F. Leakey) San

Francisco: W.H. Freeman and Company. Scientific American 239:54-73.

Ward, S.C. and D.R. Pilbeam 1983. Maxillofacial morphology of Miocene hominoids from Africa and Indo-Pakistan. In *New Interpretations of Ape and Human Ancestry* (edited by R.L. Ciochon and R.S. Corruccini). New York: Plenum Publishing Corporation. pp. 211-238.

Washburn, S.L. 1961. *Classification and Human Evolution*. Chicago: Aldine Publishing Co.

Washburn, S.L. 1976. Tools and human evolution. In *Avenues to Antiquity—Readings from Scientific American*. San Francisco: W.H. Freeman and Company. pp. 15-27.

Washburn, S.L. 1978. The evolution of man. In *Evolution. Scientific American* September issue. San Francisco: W.H. Freeman and Company. pp. 92-101.

Washburn, S.L. and E.R. McCown 1978. *Human Evolution: Biosocial Perspectives*. Menlo Park, California: The Benjamin/Cummings Publishing Co.

Washburn, S.L. and E.R. McCown *The New Science of Human Evolution*.

Weidenreich, F. 1937. The dentition of *Sinanthropus pekinensis*: a comparative odontography of the hominids. *Palaeontol. Sinica*, Peking 101:

Weidenreich, F. 1945/6: *Apes, Giants and Men*. Chicago: University of Chicago Press.

Weidenreich, F. and C.S. Coon 1965. *The Living Races of Man*. New York: Alfred A. Knopf.

Weis, K.M. 1973. Demographic models for anthropology. *American Antiquity* 38, part 2, Memoir 27.

Weis, K.M. 1975. Demographic disturbance and the use of Life Tables in Anthropology. In *Population Studies in Archaeology and Biological Anthropology* (edited by A.C. Swedlund). Society for American Archaeology pp. 46-56. (Issued as *American Antiquity* 40(2), pt. 2).

Weis, K.M. 1989. A survey of human biodemography. *Journal of Quantitative Anthropology* 1:79-151.

Weissenbach, J. 1992. A second generation linkage map of the human genome. *Nature* 359:794-801.

Werth, E. 1928. *Der Fossile Mensch*. Berlin: Bornträger.

White, T.D., D.C. Johanson and W. Kimbel 1981. *Australopithecus africanus*: its phyletic position reconsidered. *South African Journal of Science* 77:445-470.

White, T.D., G. Suwa and B. Asfaw 1994. *Australopithecus ramidus*, a new species of early hominid from Aramis, Ethiopia. *Nature* 371: 306-312.

Wilson, A.C., V.M. Sarich and L.R. Maxson 1974. The importance of gene rearrangement in evolution: evidence from studies on rates of chromosomal protein and anatomical evolution. *Proceedings of the National Academy of Sciences,* U.S.A. 71:3028-3030.

Wolpoff, M.H. 1980a. Cranial remains of middle Pleistocene European hominids. *Journal of Human Evolution* 9:339-358.

Wolpoff, M.H. 1980b. *Palaeoanthropology*. New York: Alfred A. Knopf.

Wolpoff, M.H. 1981. Allez *Neanderthal. Nature* 289:823.

Wolpoff, M.H. 1983a. Lucy's lower limbs: long enough for Lucy to be fully bipedal? *Nature* 304:59-61.

Wolpoff, M.H. 1983b. Australopithecines: the unwanted ancestors. In *Hominid Origins* (edited by K. Reichs). Washington: University Press. pp. 109-126.

Wolpoff, M.H. 1983c. *Ramapithecus* and human origins: an anthropologist's perspective of changing interpretations. In *New Interpretations of Ape and Human Ancestry* (edited by R.L. Ciochon and R.S. Corruccini). New York: Plenum Publishing Corporation. pp. 651-676.

Wolpoff, M.H. 1988. Divergence between early hominid lineages: the roles of competition and culture. In *Evolutionary History of the 'Robust' Australopithecines* (edited by F.E. Grine). New York: Aldine de Gruyter. pp. 485-497.

Wolpoff, M.H. 1996. *Human Evolution.* New York: McGraw Hill, Inc.

Wolpoff, M.H. and A. Thorne 1991. The case against Eve. *New Scientist* 22nd June. pp. 37-41.

Woo, J.K. 1962. The mandibles and dentition of *Gigantopithecus. Palaeontologia Sinica* 11: 1-94 (Quoted from Chopra, 1979).

Wood, B. 1992a. Origin and evolution of the genus *Homo. Nature* 355:783-790.

Wood, B. 1992b. Old bones match old stones. *Nature* 355:678-679.

Wood, B. 1993. Rift on the record. *Nature* 365:789-790.

Wood, B.A. and M. Collard 1999. The human genus. *Science* 284: 65-71.

Wood, B.A. and A. Turner 1995. Out of Africa and into Asia. *Nature* 378: 239-240.

Wood, B.A. and Q Xu 1991. Variation in the Lufeng dental remains. *Journal of Human Evolution* 20: 291-311.

Wright, S. 1931. Evolution in Mendelian populations. *Genetics* 16: 96-159.

Wright, S. 1969. *Evolution and the Genetics of Population.* Chicago: University of Chicago Press. Vol. II.

Wright, S. 1982. Character change, speciation and the higher taxa. *Evolution* 36:427-443.

Wu, R. and C.E. Oxnard 1983a. *Ramapithecus* and *Sivapithecus* from China: some implications for higher primate evolution. *American Journal of Physical Anthropology* 5: 303-344.

Wu, R. and C.E. Oxnard 1983b. Ramapithecines from China: evidence from tooth dimensions. *Nature* (London): 306:258-260.

Wu, R. and C.E. Oxnard 1983c. *Ramapithecus* and *Sivapithecus* from China: some implications for higher primate evolution. *American Journal of Primatology* 5: 303-344.

Wu, R. and L. Wang 1988. Single species and sexual dimorphism in *Sinoadapis. Acta Anthropologica Sinica* 7: 1-8.

Wu, X. and F.E. Poirier 1995. *Human Evolution in China. A Metric Description of the Fossils and a Review of the Sites.* New York: Oxford University Press.

Xu, Q. and Q. Lu 1979. The mandibles of *Ramapithecus* and *Sivapithecus* from Lufeng, China. *Vertebrata Palasiatica* 17:1-13.

Xu, Q., Q. Lu, J. Pan, K. Chi, C. Zhang and L. Zheng 1978. On the fossil mandible of *Ramapithecus lufengensis*. Kexue Tongbao 23: 554-556.

Yale-Cambridge expedition 1935. Refer Gregory *et al.,* 1938.

Yale-North India expedition 1931-33.

Yunis, J.J. and O. Prakash 1982. The origin of man: a chromosomal pictorial legacy. *Science* 215:1525-1530.

Zapfe, H. 1958. The skeleton of *Pliopithecus (Epipliopithecus vindobonensis)* Zapfe and Hürzeler. *American Journal of Physical Anthropology* 16:441-458.

Zorab, P.A. 1961. The historical and prehistorical background of ankylosing spondylitis. *Proc. Royal Society of Medicine* 54:415.

Index

D

Y

Z